·CONTENTS·

Introduction

Although I have almost every Sri Lankan plant book that has been published, many of them are only useful after someone has identified a tree for me. This book has been produced with people like me in mind. It does not contain photographs of all aspects of every species – it is hoped that updates will achieve this – but it should provide a good start for many people.

The aim of the book is to introduce residents and visitors to the most common trees. As it is targeted at non-botanists, the coverage is biased towards the lowlands (where most Sri Lankans live), as well as towards local and introduced species that people are likely to see planted by roads and in gardens. The trees are arranged by family, which should enable readers to start getting their heads around the relationship between plants (especially ones seemingly dissimilar) at family level, and develop the confidence to progress to more advanced books. These include *A Field Guide to the Common Trees and Shrubs of Sri Lanka*, a multi-author effort published by the Wildlife Heritage Trust, and the even more technical *Revised Handbook to the Flora of Ceylon*.

Climatic Zones & Monsoons

The topography of Sri Lanka comprises lowlands along the perimeter, which in the southern half give rise within a short distance to the central hills, rising to above 2,400m in altitude. The island can be divided into three peneplains, or steps, which were first described by the Canadian geologist Frank Dawson Adams in 1929. The lowest peneplain occurs at 0–30m, the second rises to 480m and the third rises to 1,800m.

Sri Lanka can be broadly divided into four regions (low-country wet zone, hill zone, low-country dry zone and intermediate zone), resulting from the interactions of rainfall and topography. Rainfall is affected by monsoonal changes that bring rain during two monsoons; the south-west monsoon (May–August) and the north-east monsoon (October–January). Their precipitation is heavily influenced by the central hills. These monsoons deposit rain across Sri Lanka and contribute to the demarcation of climatic regimes.

LOW-COUNTRY WET ZONE

The humid, lowland wet zone in the south-west of Sri Lanka does not have marked

seasons, being fed by both the south-west and north-east monsoons. It receives 200–500cm of rain from the south-west monsoon, and afternoon showers from the north-east monsoon. Humidity is high, rarely dropping below 97 per cent, while temperatures range from 27° C to 31° C over the year.

This is the most densely populated area in Sri Lanka. The coast is well settled, while the interior has coconut and rubber

Paddy field bordered by village gardens in Talangama.

plantations, some rice (paddy) cultivation and small industries. Remnants of rainforests and tropical moist forests exist precariously in some parts of the interior, under pressure from an expanding population. It is in these forests that most of the endemic plant and animal species that are a draw to ecotourists can be found.

HILL ZONE

The mountainous interior lies within the wet zone and rises to more than 2,400m. Rainfall is generally well distributed, except in Uva Province, which gets very little rain during June–September. Temperatures are cooler than in the lowlands, and can vary from chilly in the mornings to warm by noon. In the mid-elevations such as the area around Kandy, the temperature varies between 17° C and 31° C during the year. Temperature variations during a 24-hour cycle are, however, of a smaller magnitude. The mountains are cooler, within a band of 14–32° C during the year. There may be frost in the higher hills in December and January, when night-time temperatures fall below zero.

The central hill zone is intensely planted with tea, but has small areas of remnant forest and open grassland.

Cloud forest in Knuckles holds many endemic trees.

LOW-COUNTRY DRY ZONE

The rest of the country, three-quarters of Sri Lanka's land area, consists of the dry zone of the northern, southern and eastern plains. These regions receive 60–190cm of rain each year, mainly from the north-east monsoon. The dry zone is further divided into the arid zones of the north-west and south-east, which receive less than 60cm of rain because they are not in the direct path of the monsoonal rains.

The coastal plains in the Southern Province, where Yala and Bundala National Parks are located, and the North Central Province, home to the cultural sites, are dry and hot. Much of the dry zone is under rice and other field crops, irrigated by

Grassland and monsoon forest in Minneriya.

vast man-made lakes (called tanks, or *wewas*), many of which are centuries old, and were built by royal decree to capture the scarce rainfall in these areas. Once the 'Granary of the East', exporting rice as far as China and Burma, wars and invasions, and malaria and other diseases, laid waste to vast areas of this zone. The once-bountiful rice plains were reclaimed by scrub jungle, the haunt of elephants, bears and Leopards. Since independence in 1948, successive governments have vigorously promoted colonization and resettlement of these areas. Sandy beaches fringe the coastline and it is always possible to find a beach that is away from the path of the prevalent monsoon.

INTERMEDIATE ZONE

This is a transition zone between the dry and wet zones. Recent rainfall data shows that the wet zone with the highest precipitation is smaller than shown in maps of a few decades ago.

Habitats & Top Sites

For a moderately sized island, Sri Lanka offers a variety of habitats. It is an oceanic island with a continental shelf coming close to it in the south at Dondra, and also near the Kalpitiya Peninsula. Submarine canyons cut into the Trincomalee harbour in the north-east. A chain of partially submerged islands (Adam's Bridge) connects the island to the Indian mainland. The entire coastline offers suitable habitat for mangroves, with the western coast especially off Kalpitiya and Mannar being quite rich. The island is dotted with more than 2,000 man-made freshwater lakes in the dry zone. As many as 103 river systems create a rich network of aquatic habitats that are further enriched by paddy fields, which form an important artificial system of wetlands. The dry zone is characterized by grassland, thorn scrub and wooded sections where the soil and rainfall support these plants, or where 'gallery forests' remain along watercourses. The lowlands in the south-west originally held rainforests, but many of these have been lost to cultivation. At higher elevations, the highlands in the wet zone hold cloud forest, now a tiny remnant of what once existed. The highlands are interspersed with patana grassland.

In the context of visiting foreign botanists, the endemic plants that are found in the wet zone would be key targets. For this, a good lowland rainforest site such as Sinharaja, and a montane site such as Horton Plains National Park, are essential places to visit. Another accessible rainforest is Kithulgala Rainforest. Given that Sri Lanka is the best location for big game safaris outside Africa, a visit to a national park such as Yala is recommended – it also provides a chance to see some of the plants in the dry zone. Sri Lanka has a number of floristic regions, and many species are confined to small areas. There are also species confined to sites such as Ritigala, an inselberg in the dry zone of the northern plains, or those in areas such as Nilgala or Moneragala in the east. Isolated mountain wildernesses such as the Knuckles hold endemic species, and there are probably many species awaiting discovery. Even well-known parks and reserves such as Horton Plains and Sinharaja have had limited attention from botanists and probably hold species unknown to science.

BOTANICAL GARDENS & ARBORETUMS

The Department of National Botanical Gardens website provides a listing of sites under its purview. The main botanical gardens are the Peradeniya Royal Botanical Gardens in the mid-hills in Kandy, the Hakgala Botanic Gardens in the highlands (close to Nuwara Eliya) and the Henarathgoda Botanic Gardens in Gampaha, in the wet lowlands about an hour's drive from Colombo. These gardens are old and provide some fine examples of old trees. However, they have traditionally been somewhat ornamental in outlook and are biased towards trees from other countries rather than showcasing native species. More recently

TOP SITES TO VISIT	
Lowland Wet Zone	
Talangama Wetland	Talangama Wetland, in Colombo's suburbs, is visited by people on wildlife tours on arrival or before departure. Good range of wetland plants and trees in urban environments.
Beddegana Wetland Park & Diyasaru Park	Located within Colombo's metropolitan sprawl, these are are two fabulous urban wetland reserves with labelled trees.
Bodhinagala	Small but rich patch of rainforest about an hour and a half from Colombo.
Sinharaja	The most important site for Sri Lanka's endemic fauna. Half the trees here are endemic.
Morapitiya	Morapitiya, on the way to Sinharaja, can offer some fine views of rainforest trees. Check the state of the access road, as this is highly variable.
Kithulgala	Rainforest in the mid-hills, important because it provides an altitudinal gradient between the lowlands and highlands.
Montane Zone	
Horton Plains National Park	Important site for montane flora.
Hakgala Botanical Gardens	Good for birds and other animals, with many species habituated to people. However, most of the plants here are introduced species.
Dry Lowlands (South)	
Yala National Park	Best known for Leopards, Elephants and Sloth Bears. Also a good site for dry-zone flora.
Uda Walawe National Park	Grassland and monsoon forest. Some fine examples of dry-zone trees standing in isolation, but you cannot access the trees on foot as you may not alight from the safari vehicles.
Dry Lowlands (North-Central)	
Mannar Island, Kalpitiya Peninsula	Extensive stretches of mangroves.
Minneriya & Kaudulla National Parks	For the Elephant Gathering, which peaks in August and September. There are also some fine stands of monsoon forest and many fine specimens of trees such as Ebony.
Mid-Hills	
Kandy	Not on the itinerary for wildlife tours, but the forests around Kandy are botanically rich. In the town centre is Udawattakele, a forest park. Close by is the famous Peradeniya Botanical Gardens. A long day excursion is possible to the Knuckles Wilderness.

created botanic gardens are the Dry Zone Botanic Gardens, Mirijjawila and the Wet Zone Botanic Gardens, Avissawella. To learn about native species, especially the endemics of the wet zone, the Kottawa Forest Arboretum near Galle is one of the best places to visit, as it has a few hundred metres of footpaths with labelled trees. The arboretum in Sinharaja is just a few decades old, and the labelled trees are short in height – but this does confer the advantage of the leaves being at head height or less, making them easier to inspect.

For dry-zone plants, consider a visit to the Popham Arboretum in Kandalama, near the archaeological and cultural site of Dambulla. The shared private grounds of the Cinnamon Lodge and Chaaya Village have some fine examples of labelled dry-zone trees. This is also the case with Sigiriya Village, a large tourist hotel near the archaeological site of Sigiriya. Although these are not public gardens, you should be able to visit them to use the restaurant and enjoy the trees. Many of the leading hotel chains in Sri Lanka, such as Aitken Spence, John Keells and Jetwing, have knowledgeable naturalists who can assist botanists.

PLANT PHOTOGRAPHY

In the government botanic gardens such as Hakgala, which are used to having a lot of birders arriving with telescopes and tripods, you should not be charged for their use, though in some of the other botanic gardens there may be an extra charge. There should be no restrictions on using cameras. Colombo's Viharamahadevi Park, under the purview of the city municipality, has some fine examples of trees, though these are mainly trees of foreign import that now adorn the city streets. The park does not permit photography, on grounds of privacy for other visitors, security or restrictions on commercial use. In the national parks administered by the Department of Wildlife Conservation and the Forest Department, you are free to photograph the plants and animals you see, but even in Horton Plains, where you can go on foot, the trees are not labelled.

Sri Lankans are very friendly and outside the major cities, if you ask, people will be happy for you to even come into their gardens to photograph a plant if something catches your interest.

Plant Life

A detailed up-to-date checklist has not been included in this book, as that would be a publication in itself. *A Checklist of the Flowering Plants of Sri Lanka* by Lilamani Kumudini Senaratne (2001), included 214 families, 1,522 genera and 4,143 species. Of these, about 75 per cent were considered indigenous and about 25 per cent exotics. Of the indigenous plant species, 27.5 per cent were endemic to Sri Lanka. Of the exotics, about a third were naturalized, with the remaining two-thirds under cultivation. In the two decades since publication, a few more species have been described to science, and the number of introduced exotics has almost certainly increased. However, the figures from 2001 are broadly indicative of the species diversity in Sri Lanka.

The list of plants (and hence families) recorded in Sri Lanka continues to grow, with 217 now recognized. Additions to the list are a result of introductions of alien plants that have become naturalized, as well as the discovery of indigenous plants that have not been recorded before. Species entirely new to science continue to be added. Trimen described more than 600 plants species in Sri Lanka occurring as trees.

WHAT IS A TREE?

Plants can be grouped in various ways by the way they grow. A tree is a plant that is defined by its mode of growth. There are plants that are annuals, which grow from seed to full maturity within a year, then die. Perennials continue to grow over several years. Plants are also divided according to whether they are woody or non-woody, or herbaceous or non-herbaceous, depending on whether they have woody stems or not. Generally, annuals are herbaceous. The development of wood requires an investment in energy, and a plant with a short-lived annual cycle has too little time to invest in a woody stem.

A tree can be generalized as a woody perennial with a bole (tree trunk) taller than 6m. A shrub is a woody perennial that is multi-branched near the base. However, a plant can grow as a tree in certain physical environments and as a shrub in others. All plants are influenced by environmental factors, a field of study known as epigenetics. Although individual plants of the same species may have the same DNA, how the genes in the DNA execute the molecular code embedded in them will vary in response to the physical environment they are in. Take, for example, the seeds from the same parent plant, some of which are growing in a damp, shaded area, while others, only a few hundred metres away, are planted in dry, warm gravel. The offspring plants can look like two unrelated species.

THE PARTS OF A FLOWER

In a flower, the male part comprises the stamen, which has pollen-bearing anthers on top of a filament. The female part consists of a stigma, which is a pollen receptor, and a tube called the style that leads into the ovary. The ovary contains one or more ovules that when fertilized develop into seeds. When pollen lands on the stigma, it grows a thin tube along the style to the ovary to fuse with the female cells (ovules), and starts to develop as a seed. In botanical terms, the fruit is the enlarged ovary containing the seed or seeds, but more generally the term 'fruit' refers to the whole structure containing the seeds, which can include a fleshy or hard covering around the seed-bearing ovary.

The ovary, style and stigma are collectively known as a carpel. If a flower has only one carpel or has free carpels, these individual carpels are referred to as a pistil. Where a flower has two or more fused carpels (syncarpous), the term pistil refers to all of them. Below the petals are small green parts; the sepals that cover the flower as a bud. In some flowers there is no clear distinction between the sepals and petals, and these are termed tepals. The sepals and petals are known as the perianth. If the ovary is below the perianth it is known as inferior and if above, it is superior. In some families (like the capers), the ovary is raised well above the perianth on a stalk known as a gynophore.

Dioecious plants are those with separate male and female plants (that is, a plant will only contain female flowers or male flowers). This makes cross-pollination easy as there is no risk of self-pollination. Monoecious plants have both male and female flowers, or bisexual flowers (containing both male and female parts). One strategy to avoid self-pollination is to have the male and female parts mature at different times. If both male and female parts mature at the same time, plants wishing to avoid self-pollination have to develop other mechanisms. However, some plants elect for self-pollination. This can happen, for example, in environments where pollinators are absent.

The ovary contains the ovules, which after fertilization develop into an embryo or young plant neatly folded within a seed. The fruits of flowering plants (angiosperms) can take many forms. In popular parlance the term 'fruit' is used in cases where there is an edible, fleshy covering. However, in botanical terms a nut is also a fruit. A fruit such as the pineapple is actually a compound fruit made up of many fruits.

The edible figs produce what are popularly referred to as fruits. But the fleshy part that we and other animals eat is actually a receptacle upon which (on the inside of the fig) tiny flowers grow out of human sight. There is a tiny hole in the fig through which pollinating fig wasps (see p. 91) enter. The fertilized ovules of the flowers develop into seeds and the normal cycle from ovules to seeds occurs hidden from view. Carpels are often contained within a fleshy covering that is perceived as a fruit. In many plants, the carpels split (or dehisce) into the component carpels.

POLLINATION & DISPERSAL

For the next generation of plants to be produced through sexual reproduction, the ovules must be fertilized. As mentioned, this requires pollen from stamens to be deposited on the stigma. Plants achieve pollination using three main vectors: wind, water and animals.

- For pollination by wind, the pollen must be very light and produced in large quantities to achieve results.
- To use animals, the plants have to provide food, usually in the form of nectar, to offer the animals an inducement to visit the flowers and to unwittingly carry the pollen to another flower. This has given rise to many complex and intricate relationships between plants and their pollinators through co-evolution.
- Pollination by water happens in two ways. Some aquatic plants produce light pollens that float through the water's surface and reach a stigma. Some other species, especially submerged ones, produce heavier pollens that sink and are caught by the stigma.

Plants also use the three elements of wind, water and animals to disperse their seeds so that they may find the right physical conditions in which to grow into a new plant. Once again, with animals, the plants have to offer a reward, and this is in the form of energy-rich food covering the seed or the seeds themselves being edible.

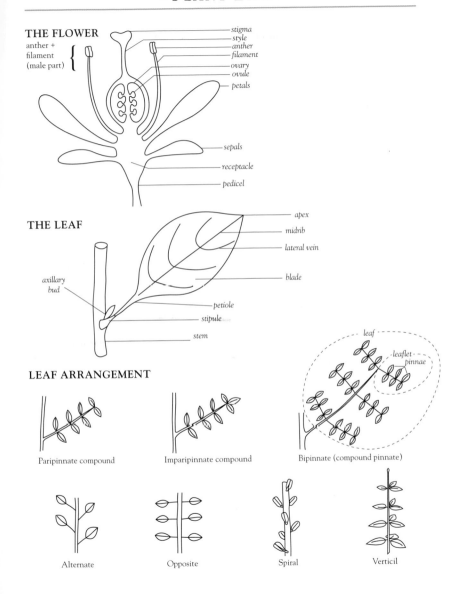

THE FLOWER

anther +
filament
(male part) {

- stigma
- style
- anther
- filament
- ovary
- ovule
- petals
- sepals
- receptacle
- pedicel

THE LEAF

- apex
- midrib
- lateral vein
- blade
- petiole
- stipule
- stem

axillary
bud

leaf
leaflet
pinnae

LEAF ARRANGEMENT

Paripinnate compound

Imparipinnate compound

Bipinnate (compound pinnate)

Alternate

Opposite

Spiral

Verticil

INFLORESCENCE TYPES

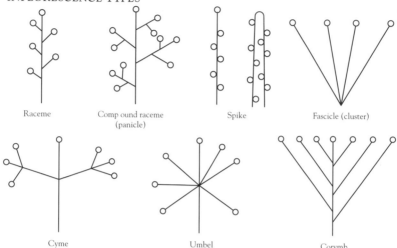

Raceme

Comp ound raceme (panicle)

Spike

Fascicle (cluster)

Cyme

Umbel

Corymb

FLOWER ARRANGEMENT

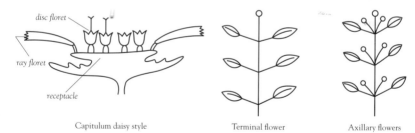

disc floret

ray floret

receptacle

Capitulum daisy style

Terminal flower

Axillary flowers

LEAF SHAPES

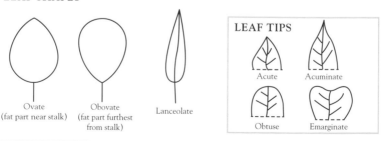

Ovate
(fat part near stalk)

Obovate
(fat part furthest
from stalk)

Lanceolate

LEAF TIPS

Acute

Acuminate

Obtuse

Emarginate

ARRANGEMENT OF TREES

In this book the trees and shrubs are arranged by family because the use of photographs showing more than one aspect of a tree does make it easier now to identify trees arranged in this way than was the case a few decades ago, when only hand-drawn illustrations were used. The big advantage of this is that it helps people to get started at being better amateur botanists or naturalists as they begin to learn the characteristics that define a family. Furthermore, in the tropics, the different forms that species of a family can take are so varied that keeping them together makes it easier for people to learn the relationships.

Taxonomic Classification

Identifying a species in a group, whether it is birds, mammals, butterflies or dragonflies, becomes easier when you become familiar with what 'type' of animal it is. Scientists group all species into taxonomic hierarchies. A simple hierarchy would be where related species are grouped into genera, which are in turn grouped into a family, and in turn into an order. The tree of life attempts to explain the inter-relationship of species from an evolutionary point of view.

The number of levels or taxonomic ranks (like families and orders) used can vary for different groups of living things. Different authorities may have different views on the levels in the hierarchy. The various ways in which scientists attempt to group living things is the science of taxonomy. Controversy and debate rage on this subject, and as a result often different books classify species differently. Classification is also seemingly complicated with the use of suborders, subfamiles and superfamilies, with which different authors arrange species into different hierarchies. Animals such as butterflies become even more complicated with the use of tribes. But these are all ways of trying to map species into a tree of life to understand how they are related to each other through the act of evolution over time. Most people can ignore the classification hierarchies and be content with simply getting a feel for where things approximately belong in terms of simpler relationships at family and order level. In this book, the trees are grouped simply, under families, and a brief description of each family is included within the species accounts (see also p. 15).

The common name of a species is followed by its Latin name, which generally has two parts, the genus and the specific epithet. No two species can have the same Latin binomial at a time, and the Latin names are thus relatively stable. Common names, on the other hand, can vary widely from one country to another or even within regions in a country. Some species will have a trinomial, described as a subspecies or geographical race. For example, a mammal species, the endemic Purple-faced Leaf Monkey in Sri Lanka, has four distinct subspecies. It logically follows that all four subspecies are endemic. In the case of dragonflies, the animal found in Sri Lanka, known as the Sri Lanka Cascader *Zygonix iris ceylonicus*, is found only as a single subspecies, which is endemic. It has subspecies status to distinguish it from other populations found elsewhere in the world similar enough to be considered the same species but different enough to be treated as distinct subspecies.

Plants have many varieties, usually when they have been bred for horticulture. However, wild plants can also have naturally occurring varieties or subspecies. Recognizing subspecies in plants is very tricky unless there are clear geographical boundaries. This is because the influence of epigenetics, where the environment interacts with the instructions in DNA, is more obvious in plants than in animals. Plants of a specific species grown from seeds that take root in open, sunny and poor soil can look very different from plants of the same species growing from seeds that have taken root tens of metres away under shade in nutrient-rich soil – they may look like two different species. A particular plant may grow as a shrub on an exposed mountaintop and as a tree in a more sheltered environment.

By way of explanation of the way classification works, the taxonomic hierarchy for the Ceylon Ironwood looks like this:

Kingdom	Plantae
Division	Angiosperms
Unranked clade	Eudicots
Order	Ericales
Family	Sapotaceae
Genus	*Manilkara*
Species	*Manilkara hexandra*

What are known as seed-bearing plants have two natural divisions, into gymnosperms and angiosperms. Traditionally, the flowering plants were split into monocotyledons and dicotyledons. Monocotyledons are plants with one seed leaf in the embryonic plant and are plants such as grasses, in which the leaves are thin and long and have parallel veins.

The dicotyledons with two seed leaves in the embryonic plant are also known as broadleaved plants and are the rest of the seed plants. This division is, however, not a natural grouping, as it is not a result of a simple, two-branched evolutionary split. Phylogenetic analyses show that ancient dicotyledonous plants split into monocotyledons and into a further branch of new dicotyledons. Thus the flowering plants we see today are made up of a mix of old dicots (like magnolias), monocots and newer dicots (like sunflowers). In taxonomic terms the monocots are monophyletic, meaning that everything classified as a monocot has a single common ancestor from the events that led to the older dicots branching into monocots and more modern dicots.

Species Descriptions

The species accounts have been written with a view to help people identify trees in the field. Commercial and medicinal uses are not covered in detail, as these are given in many other sources. The text focuses on helping readers to put a name to a tree. There are gaps in the accompanying images used to illustrate the species, and I would be happy to hear from anyone who would like to share images to make future editions more comprehensive (see p. 176).

The tree descriptions have been written to be intelligible to the non-botanist and they often first describe a feature in plain English, then provide the botanical term in parentheses. The objective is to get people interested in botany and not to put them off by making them feel as though they have to learn a new language first. For more detailed books, see Bibliography (p. 168).

FAMILY ACCOUNTS

The species are grouped under families. Many of the most common species in Sri Lanka are introduced and have been naturalized for centuries in Sri Lanka, so that most people are not aware of their foreign origin. The family descriptions give a context to the trees' worldwide distributions (providing family distributions at a country level is not generally very meaningful, as many families include species that may be adapted to different environments, from the lowlands to the highlands).

ESSENTIAL PLANT TERMS

As already mentioned, technical terms have been kept to a bare minimum, though there are a few terms that the reader will need to become familiar with as they are too fundamental to be repeatedly explained. They include: alternate, axil, calyx, cyme, drupe, opposite, ovary, panicle, raceme, sepal, stamen, stipule, style and stigma (see Glossary, p. 16).

SPECIES ACCOUNTS

The species descriptions use a standard structure with the English and scientific name, followed by the Sinhala (S) and Tamil (T) names where known. The text covers the structure, flower, leaf, fruit, habitat and distribution of each plant. The etymology of species is included where possible, and explanations of the origins of the scientific names help to demystify the plants.

GLOSSARY

Technical terms have been avoided as much as possible in the text. However, at times – to avoid long-winded explanations – standard botanical terms have been used. These are included below, along with a number of other standard botanical terms that are useful to know when consulting more advanced texts.

achene Single-seeded fruit, dry and with a thin wall, not splitting when ripe.

actinomorphic Radially symmetrical. Flower can be divided into two mirror-image halves by dividing through circumference.

acuminate Tapering to long tip (e.g. leaf-tips).

adaxial Refers to side of an organ that bears in direction of axis to which it is attached. The upper surface of a leaf is adaxial to the stem; the under surface is abaxial.

adnate Attached to or pressed against. Usually refers to different structures or organs in a plant, e.g. petals adnate to staminal tube.

adventitious roots Roots arising from somewhere other than the root system.

aerial roots Roots arising from above the ground. In the case of some mangrove plants these are in addition to the main roots, which are underground. Epiphytic plants also have aerial roots that dangle and are not connected to soil.

alternate In relation to leaves, alternate leaves are spaced apart on opposite sides and not directly opposite each other.

amplexicaul Where a leaf or stipule has its base growing around to the opposite side of the stem to embrace it.

androecium Male parts of a flower comprising stamens in a flower. Each stamen consists of an anther and filament.

androgynophore Stalk carrying both stamens and ovary, arising from above where the petals are joined (e.g. passion flowers, *Passiflora* spp.).

anisophylly Pair of opposite leaves at a node with a pronounced difference in size or shape.

apocarpous Where carpels in a flower are free. If they are joined or fused together, they are syncarpous.

appressed Lying flat and close to stem, for example in the case of hairs or branches. Can also refer to being pressed against the ground for plants that are appressed to the soil (e.g. sundews).

auricle Ear-like projection at base of leaf, leaf blade or bract.

axil Angle between stem and leaf stalk.

axillary Arising on axil, as in the case of stipules or flowers, between angle of leaf stalk and stem.

basifixed In context of anthers, refers to anthers fixed at the base to the filaments.

bifid Split in two, e.g. apex of a leaf, split or divided into two.

bisexual Having both male and female parts in the same flower.

bole Unbranched part of a tree trunk (above buttress if one is present).

bract Small modified leaf at base of an inflorescence or flower stalk.

bracteole Small modified leaf below flower or between bract and flower.

bulb Underground stem, as in onions.

caducous Tendency of leaves (including stipules) to fall off easily or prematurely. Not persistent.

calyx Outermost whorl of a flower's organs, often divided into sepals.

capillary Slender, nearly hair-like.

capitulum Densely packed head of stalkless

flowers on a flat base (e.g. daisy).

cataphyll Refers to scale leaf or scale-like leaf.

caudate Refers to leaves with pronounced drip tip. Leaf shape ends abruptly with extended thin tip.

chartaceous Thin and stiff; like paper.

ciliolate Fringed with very small hairs (e.g. calyx of *Derris parviflora*).

circinate Coiled inwards, as in circinate tendrils.

clade In taxonomic terms, group comprising its ancestor and its descendants.

clavate Club shaped.

connate Joined or attached to. Refers to similar parts, e.g. connate petals that may be fused at base.

cordate Leaf that is deeply notched at base to form a heart shape.

coriaceous Refers to leaves that are stiff or tough; also at times described as leathery. Subcoriaceous refers to leaves that are fairly stiff and tough.

corm Underground swollen stem used for storage (e.g. in *Colocasia* spp.).

corolla Second whorl of a flower's organs, inside or above calyx, and outside stamens. Corolla is formed of either fused or separate petals.

corona This is formed by a series of appendages on the petals of the corolla, the backs of the stamens, or at the junction of the corolla tube and petals. The appendages are often united to form a ring (e.g. Passifloraceae). See also Amazon Lily *Eucharis grandiflora*.

corymb Type of racemose inflorescence where flowers form a flat top. This means that the inner flowers (the higher ones) are on shorter stalks than the outermost (lower) ones.

crenate Crenate leaf margin is notched with regular, rounded, symmetrical teeth.

crenulate Margins with small, rounded, symmetrical teeth.

culm Stem of grass or sedge.

cyme Inflorescence where central flower is the oldest. Later flowers arise from leaf axils in central flower. Other flowers arise from leaf axils in these. Can be a simple or compound cyme. A thyrse is where a raceme has flowers replaced by cymes.

deciduous Refers to plant that seasonally sheds leaves. In high latitudes, plants shed leaves in winter and in the tropics during dry periods.

decurrent When leaf blade runs down leaf stalk and it is difficult to separate the two; also used when leaf blade forms wing around petiole or stalk.

decussate Alternate; opposite pairs of leaves at right angles to each other.

dehiscent Splitting or dehiscing along predetermined lines when mature; applied to fruits and anthers.

deltoid, deltate Shaped like equal-sided triangle. Term deltate is preferred. Mostly applied to shapes of leaves.

dentate Toothed margin (typically of leaf) with prominent, symmetrical, sharp points.

denticulate Finely toothed. A denticle is a fine tooth.

depressed Flattened from above downwards (e.g. fruit of Cherry Guava).

dichasial cyme Where an inflorescence has a central stalk with a terminal flower, and laterally and opposite to each other are two branches with a terminal flower, and again each of these has a pair of lateral and opposite branches bearing flowers.

dicotyledon Plants with two seed leaves in embryo.

digitate Refers to compound leaf where

individual leaflets originate from same point. Also known as palmate in reference to fingers originating from palm.

dioecious Refers to plants with unisexual flowers, with male flowers and female flowers being on separate plants, so each will be either a female or male plant.

disc floret In the Asteraceae (daisy family), a flower head comprises disc florets in the central disc, and on the rim are ray florets that look like petals.

distal The further end from the point of attachment. Distal end of a leaf will be leaf tip (as opposed to leaf base).

distichous Arranged in two vertical rows.

domatia Structures, usually in leaf, but also in stem or root, inhabited by animals, especially ants.

drupe Fleshy, indehiscent (not splitting) fruit with seed having hard or stony covering (endocarp) surrounded by fleshy pulp (e.g. cherry).

emarginate Notched at tip, e.g. of a leaf.

ensiform Long and narrow leaves, ending with a point (e.g. pineapple leaves).

entire In context of leaves, refers to undivided margin. Leaf margin is smooth without crenulations, lobes and so on.

epicalyx Whorl of bracts beneath flower that looks like a second calyx. Common feature in shoe flowers.

epipetalous Usually refers to stamens that are united with the petals and appear to lie on them.

ericoid With needle-like leaves, typical of heathland plants.

exserted Projecting as from a sheath or pod.

exstipulate Without stipules.

fascicles Cluster of similar organs all arising from same point on plant, e.g. cluster of leaves, fruits or stamens.

flower Reproductive organ of a plant. Part of plant with male or female parts or both capable of reproduction. What looks like a flower to a layperson may in the botanist's eye be a collection of flowers and be more correctly an inflorescence.

foliolate With leaflets.

follicle Dry fruit formed from single carpel that splits along one side (axis) only.

fruit Strictly speaking, the fertilized, mature ovary of a seed-bearing plant. More widely used to include berries, drupes and compound fruits.

funicle Stalk of ovule attaching it to placenta.

fusiform Spindle shaped, thick in middle and tapering at both ends.

gamopetalous Petals fused, at least at base.

gamosepalous Sepals fused, at least at base.

glabrous Smooth, that is without hairs or other epidermal growths. A glabrous leaf surface is smooth and shiny without hairs.

gregarious Refers to plants occurring in communities, e.g. dipterocarp communities in lowland rainforests.

gynodioecious Plants that bear either female or bisexual flowers.

gynoecium Female parts of flower comprising one or more fused or free carpels. Each carpel contains an ovary, style and stigma.

gynophore In some families (e.g. capers), the ovary is raised well above the perianth on a stalk known as a gynophore.

halophytes Plants adapted to salty conditions.

hastate Where base of a leaf comprises two approximately triangular projections or lobes that are outwards pointing. Hastate leaves can be arrow shaped.

herbaceous Refers to herb-like, that is non-woody plant.

hermaphrodite Refers to bisexual flowers, where male and female parts are found in the same flower.

heterostyly Plants that have flowers with two or more different lengths of style; an adaptation to improve pollination success.

hilum Scar left on a seed at the point at which it was attached to placenta.

hirsute With coarse and rather stiff hairs.

hispid With long, stiff hairs or bristles. More bristly than hirsute.

hypanthium Cup-shaped extension of basal part of flower holding sepals, petals and stamens. Sometimes incorrectly referred to as floral tube.

imparipinnate Pinnate with a single terminal leaflet. Thus, there is an odd number of leaflets.

indehiscent Refers to fruit that does not split naturally on ripening.

indigenous Occurring naturally in a country or region without having been introduced by humans.

indumentum Covering of hair or scales.

inferior ovary Sepals and petals are known as the perianth. If the ovary is below the perianth it is known as inferior and if above, it is superior.

inflorescence Collection of flowers, where a flower is the part of a plant that has male or female parts or both capable of reproduction.

infundibuliform Funnel shaped; applies especially to styles.

involucre Whorl of bracts beneath an inflorescence.

keel petal In the pea (Fabaceae) family, one large, showy petal is turned upwards and is the keel. Other petals are fused to form inverted boat-shaped keel (the standard), which encloses stamens and style to form a two-lipped flower.

labellum Lowest petal in an orchid, different from other two lateral petals and often larger.

lancelolate Ovate or egg shaped at base and narrowing at upper half to a point.

lenticels Spots on bark; may be circular or oval shaped. Lenticels may be loosely arranged into rings around bark. Cherry trees grown in parks have pronounced lenticels.

loculicidal As in a seed-containing capsule that splits longitudinally along dorsal sutures of wall.

merous Trimerous, tetramerous, pentamerous and so on. Refers to main parts of a flower (sepals and petals) appearing in groups of three, four, five and so on. In some flowers, the sepals, for example, may be trimerous and the petals pentamerous.

moniliform Root that looks like a string of beads.

monocotyledon Plants such as grasses with one seed leaf in the embryo. In adult plants leaves have parallel veins and flowers are trimerous.

monoecious Having male and female flowers or bisexual flowers in one plant.

monopodial Plant with single axis of vertical growth from a main stem, as in tall trees that have a single main stem.

monotypic Containing only one species in a genus or only one genus in a family.

mucilage Slimy excretion that swells on absorbing water.

mucronate Refers to leaf that ends with short, stiff point or tip.

naked flowers Flowers without calyx or corolla.

ob- (prefix) Has two meanings, the first being against. The second, used in botany, indicates that the attachment point is opposite to the usual shape. For example, obovate means opposite of ovate, that is the widest part of the leaf is in the apical rather than basal half.

oblanceolate Narrow, parallel-sided leaf, rounded at top and base.

opposite Refers typically to leaf arrangement where leaves are directly opposite each other on a stem, or in a compound leaf where leaflets are directly opposite each other on an axis. Alternate leaves are where the leaves are spaced apart on opposite sides and not directly opposite each other.

orbicular Flat and circular in shape; a sphere will be described as globose.

pachycauln Thick stemmed and sparsely branched.

palmatifid When a leaf has a palm-like (digitate) appearance with lobes formed with the divisions of the lobes reaching midway into the leaf.

panicle Can be thought of as a compound raceme. An inflorescence where the main stalk has other branches (which can in turn be branched), carrying flowers in a raceme (e.g. flowers of grasses).

papillae Soft, small protuberances that extend beyond or above surface.

pappus Series of bristles or hairs on base of corolla, which are later found at tip of a fruit (e.g. family Asteraceae).

paripinnate Pinnate leaflets without a terminal leaflet; even number of leaflets in a compound leaf.

patana, patna Grassy hillside.

pedicel Flower stalk of individual flower in an inflorescence.

pedicellate Stalked flowers.

peduncle Multiple meanings. In an inflorescence, lower unbranched part

of stalk is called a peduncle, and upper sections with branches of flowers the rachis. Also used to refer to stalk of a single flower or common stalk of flowers without stalks (that is, sessile flowers).

peltate Round and attached in centre, e.g. leaves of a water lily where petiole is not attached to blade margin but in centre.

pendulous Hanging, like a pendulum.

perfect Perfect flowers have both male and female parts.

perianth Floral envelope comprising outer whorl of calyx (made of sepals) and inner whorl of corolla (made up of petals). One or both whorls may fuse to form a tube.

pericarp Term that includes epicarp (skin), mesocarp (middle) and endocarp (inside) layers of a fleshy covering around a seed, in a single-seeded fruit like a mango.

persistent In the case of stipules, when they do not fall off easily after the new leaves have grown. Generally refers to parts of a plant that do not fall off as easily as expected.

petaloid Like a petal, e.g. in the case of sepals that are coloured and shaped like a petal.

petiole Stalk of a leaf.

petiolules Stalks of leaflets in a compound leaf that has component leaflets.

phloem Plant tissue used for transporting food material. In woody bark, phloem will be inside. Note that the xylem is the tissue that transports fluid.

pilose Hairy with short, thin hairs.

pinnae In this book, the first order division on a leaf midrib into distinct components (that is, not lobes) is termed leaflets. A second order division of leaflets will comprise pinnae (singular, pinna). See diagram, p. 11.

pistil Where carpels are free (or

apocarpous), or fruit has a single carpel, refers to female reproductive part of a flower comprising stigma, style and ovary. In this situation, carpel and pistil refer to the same thing. Where carpels are fused (or syncarpous), the term pistil refers to all female organs in flower. Another term for this is the gynoecium.

pollinia Club-shaped masses of sticky pollen found in orchids.

polygamous Having male, female and bisexual flowers on same plant.

prostrate Lying flat along the ground.

protandrous Stamens (male parts) in bisexual flower mature before female parts are receptive. This avoids self-pollination.

protogynous Stigmas (female receptors) in bisexual flower mature before male parts are ready to disperse pollen. This avoids self-pollination.

psammophytes Plants adapted to growing in sand or sandy soils.

pseudobulb 'Above-ground' storage organ found in many species of epiphytic and sympodial orchids.

pubescent Hairy. A pubescent leaf is one covered with downy hairs.

pulvinate Swelling on petioles either at join with stem or at join with leaf blade or both. Dipterocarps, for example, have pulvinate petioles.

pulvinus Swollen or enlarged section of leaf stalk that allows plants to change the direction (sleep movements) of their leaves in response to light.

pyriform Pear shaped.

raceme Where flowers are attached by pedicels to a single stalk. Similar to flowering spike, but flowers are on individual stalks that come off a central stalk. There are many variations to inflorescences arranged in a raceme. In a spadix and spike, flowers are sessile (without a stalk). In a raceme, newest flowers are at top and are last to open.

rachis In a compound leaf, part of central axis, between first set of leaflets and tip, excluding part of axis up to first set of leaflets, which is the petiole. Midrib or main nerve of a compound leaf. In an inflorescence, rachis is similarly the part of the axis except the first part (peduncle) up to the first set of flowers.

ray floret In the Asteraceae (daisy family), a flower head comprises disc florets in the central disc, and on the rim are ray florets that look like petals.

resupinate Flowers that are upside down (e.g. orchids) or seemingly so.

reticulate Lace-like network of veins on a leaf.

rhizome Underground stem. Can be told apart from roots by presence of nodes, buds or scale-like leaves.

rotund Nearly round (or circular). In context of leaves, assumes a nearly circular two-dimensional shape.

samara Dry, non-splitting (indehiscent) fruit with a wing, which is longer than part with the seed.

samaroid Resembles samara, but with wing surrounding seed chamber.

saprophyte Refers to plants that lack chlorophyll and cannot manufacture their own food. They need to absorb nutrients from decaying organic matter. Some orchids are saprophytic.

scales Modified leaves that protect other parts in a plant, such as the growing point, from frost. Less common in warm tropics than in cooler regions.

scape Leafless flower stalk or inflorescence stalk.

scarious Having a dried up appearance.

sepaloid When petals look like sepals (and

are functioning as sepals).

septicidal Septicidal dehiscence is where a ripe (mature) capsule splits along the lines of the carpel junctions, but the valves (parts of carpel that open out) do not fall off and remain attached. In septifragal dehiscence, the valves fall off and an axis (columella) to which the seeds are attached remains.

serrate Saw toothed, with regular acute-angled teeth (pointing towards apex).

serrulate As in serrate but with minute teeth.

sessile For example, leaves without a petiole would be sessile leaves.

shrub Woody plant that is much branched near base (a tree grows up a single, straight trunk to some height before branching).

simple In the context of leaves, simple leaves are those not divided into leaflets.

spadix Unbranched inflorescence with thick and fleshy axis. Flowers are attached to it and may in some cases be partially sunken into it.

spathe Large, enveloping, leaf-like structure enclosing cluster of flowers.

spatulate Spoon shaped; also spelt spathulate.

staminate Staminate flowers are those bearing stamens, that is male flowers. Bisexual or perfect flowers have both male and female parts.

staminode Sterile stamen, usually smaller than fertile stamen and not bearing pollen.

standard petal In pea family, one large, showy petal is turned upwards and is the keel. Other petals are fused to form an inverted boat-shaped keel (the standard), which encloses the stamens and style to form a two-lipped flower.

stipel Stipule-like outgrowths occurring at base of leaflet or pair of leaflets in compound leaves.

stipitate Supported on a special stalk, not on a peduncle, pedicel or petiole.

stipule Growth at base of leaf on leaf petiole, usually in pairs, which can be leaf-like or spine-like. Note that exstipulate means without stipules.

stolon Vegetative shoot that spreads along the ground, giving rise to new plantlets at nodes.

sub (prefix) Two meanings. The first is almost or nearly, e.g. subdeltate is nearly deltate (triangular). The second is below or under.

subulate Awl shaped, like stout needle tapering to fine point.

superior ovary Sepals and petals are known as the perianth. If ovary is below perianth, it is known as inferior and if above, it is superior.

sympodial Without a single main stem. Sympodial orchids, for example, do not have a single main stem. Sympodial growth is when growth takes place from axillary shoots or axillary branches, thus forming a discontinuous axis of growth. Flower cymes form when a central flower opens first and growth is continued with lateral flowers.

syncarpous (flowers) Flower in which carpels are fused or joined together (or united).

talawa Grassland with scattering of trees.

tank Man-made lake or irrigation reservoir.

terete Circular in cross-section.

thyrse Inflorescence where raceme has flowers replaced by cymes. A thyrse is thus a compound raceme.

tomentose With dense covering of short, soft hairs.

tomentum Covering of downy hairs, like felt.

trichome Hair-like outgrowth.

trifid Split in three.

trifoliate With three leaflets.

tristichous Leaves arranged vertically in three rows.

tuber Underground modified stem or root, used as storage organ.

tubercular, tuberculate Covered with wart-like protuberances; knobbly.

turbinate Shaped like a spinning top with top obconical.

umbel Where stalks of individual flowers are joined together at their bases.

undulate Wavy, usually referring to leaf margin.

unisexual Refers to flower that has only male parts or female parts.

urceolate Urn shaped, constricted at top and expanded again slightly to form narrow rim

valvular dehiscence When fruit splits open along sutures of carpels.

velutinous Velvety, soft to the touch.

vermiform Worm shaped, thick and bent in places; especially of roots.

verrucose Having warts or little excrescences on surface, e.g. a bark that is verrucose.

verticils Arrangement in a whorl of structures that is not typically found in whorls, e.g. flowers of Ceylon Slitwort are arranged in whorls or verticils.

villous, villose With long, soft hairs.

viscid Sticky.

viscidum Glands to which pollinia in orchids are attached.

viviparous Bearing live young, a condition common in mammals, less rarely in other animals. In plants, refers to seeds germinating on parent plants, as in the Rhizophoraceae.

whorl Branches or leaves arising from same point on stem, like spokes on a wheel. A leaf whorl occurs when there are more than two leaves together at the same level.

xylem Woody tissue used for transporting fluids in vascular plants. Note that the phloem is the tissue that transports food.

zygomorphic Bilateral symmetry; only a single vertical dividing line can be drawn, so that the flower is divided into two vertical halves, each the mirror image of the other (e.g. orchids).

Cycadaceae (Sago)

The sago family includes just one genus, *Cycas*, with a little over 100 species. Although cycads are known from ancient fossils, molecular dating suggests that the *Cycas* genus is no more than 5 million years old. Cycads are native to tropical East Africa, Madagascar, India, Sri Lanka, Australia and Polynesia. Their leaves are arranged in a whorl at the top of the stem. The leaf has arrow leaflets that are arranged pinnately on either side of a strong midrib, creating a feathery appearance to the leaves. The family has separate male and female flowers. The male flowers (sporophylls, or fertile leaves) are in a cone at the top of the stem, and the growing stem eventually thrusts the cone aside. The female flowers are in the form of an ovule bearing scales that initially form a cone, then spread out as the stem grows through the cone. These small trees look similar to palms (family Palmaceae). Their stems are typically not branched. They were at their peak in the Jurassic period, which is also known as the Age of the Cycads.

Cycad ■ *Cycas circinalis*
(S: Madu)

STRUCTURE Short, palm-like plant. Starch from the pith is used to make sago. However, the toxins have to be leached out and all parts of the plant are poisonous to humans. **FLOWER** Male flowers form in cones. Female flowers are ovules bearing scales, and are also formed within a cone. **LEAF** Eighty to 100 pairs of narrow leaflets pinnately arranged on a midrib. **FRUIT** Egg-shaped seeds reddish-yellow. **HABITAT & DISTRIBUTION** Mainly a tree of the intermediate zone. Occupies scrub forest and savannah. National parks such as Wasgomuwa have naturally occurring trees. Also widely planted in gardens as an ornamental. **ETYMOLOGY** *Cycas* is derived from a Greek word for a kind of palm. The emergent leaves are circinate – referring to a coiled arrangement like a watch spring.

MYRISTICACEAE (NUTMEGS)

This commercially important tree family is found worldwide in the tropics, and comprises about 520 species in around 21 genera. A number of canopy trees in rainforests belong to the family. Their fruit is a fleshy or woody capsule that typically splits in half. The seeds are covered in a lacy or leathery aril. The Nutmeg *Myristica fragrans* is the best-known member of the family, and is grown all over the tropics, including Sri Lanka, for its use in cuisine. It is a native of the Banda Islands, over which successive European nations fought to control the spice trade. The leaves are simple and alternate. The flowers are in racemes or panicles, and are small and radially symmetrical (actinomorphic). Separate petals and sepals are missing and instead the flowers have tepals, which are quite often fleshy.

Iriya ▪ *Horsfieldia iryaghedhi*
(S: Iriya)

STRUCTURE Medium-sized tree growing to about 25m in height. Bark greyish-brown, and generally smooth but can be a little flaky. Trunk often develops buttresses, perhaps as an adaptation to soft soils, since it often grows near water and other waterside trees like the Kumbuk (see p. 100), which develop prominent buttresses. Drooping branches. **FLOWER** Tiny orange-yellow male flowers form a flowering spray about 10cm long. Female flowers a little bigger. Flowers have a citron-like fragrance and are used as temple offerings. **LEAF** Oblong leaves tipped with a point, but not sharply pointed. Clear lateral veins from midrib. Leaves have distinct stalks. They are arranged in two ranks alternately and droop. **FRUIT** Round fruit brown on the outside, pink on the inside. Aril flame-red. **HABITAT & DISTRIBUTION** Swampy areas in the wet lowlands and rainforest sub-canopies. **ETYMOLOGY** Genus name honours T. Horsefield (1773–1859), a naturalist and doctor in the East Indies. 'Iriya' is the Sinhala name that is used as the scientific name.

Malabodha Nutmeg ■ *Myristica dactyloides*
(S: Malabodha, Sadhikka)

STRUCTURE Tall, much branched, large tree. Bark grey with hint of orange. Can look very dark in low light in a forest. Drooping branches and leaves. **FLOWER** Female flowers nearly stalkless (sessile), orange-yellow and clustered at leaf axils. **LEAF** Dark green oval leaves, with pronounced midrib. Many parallel lateral veins. Leaves leathery, and arranged alternately. **FRUIT** Egg-shaped fruit has hairy brown coat. Shiny brown seed has orange aril. **HABITAT & DISTRIBUTION** Intermediate and wet-zone rainforests. Endemic to Sri Lanka. **ETYMOLOGY** *Myristica* is derived from the Greek word *muron*, a reference to a sweet juice from plants.

ANNONACEAE (SWEETSOPS & SOURSOPS)
This family is found mainly in the tropics and subtropics, with a few species occurring in the temperate zone. It comprises primarily tropical trees, but also shrubs and lianas. There are about 2,500 species in around 135 genera. The highest number of species is found in the Old World, followed by the New World and Africa. Plants in this family are commercially important for their edible fruits, including cherimoyas, sweet sops and soursops. Their simple leaves are usually alternate in two ranks. The radially symmetric flowers are aromatic and often pendulous.

Wild Anodha ■ *Annona glabra*
(S: Wal Anodha, Wel Aatha)

STRUCTURE Small, bush-like tree that branches very low. Much branched and forms roughly spherical outline with its branches. **FLOWER** Solitary yellow flowers red inside,

with six petals. **LEAF** Oval to elliptic leaves, tapering to form blunt tip. Pronounced lateral veins on strongly marked midrib. Reticulate venation formed by tertiary veins. **FRUIT** Shiny, oval green fruit. Pulp yellow and seeds pale reddish-brown. **HABITAT & DISTRIBUTION** Common in swampy areas in lowlands, especially in the wet lowlands. **ETYMOLOGY** Genus name, *Annona*, is derived from a South American plant name.

Soursop ■ *Annona muricata*
(S: Katu Aatha)

STRUCTURE Small evergreen tree with branches sloping upwards at steep angle. Bark smooth and greyish-brown. **FLOWER** Six thick, pale yellow petals. **LEAF** Oblong leaves with pronounced but short, sharp point. Leaves aromatic. **FRUIT** Green ovoid fruit can be 20cm long, large in relation to other species in the genus. Skin dull green with prickles. **HABITAT & DISTRIBUTION** Tropical American tree widely planted in home gardens for its fruits. **ETYMOLOGY** *Muricata* is a Latin reference to being armed with points.

Sweet Sop or Sugar Apple ■ *Annona squamosa*
(S: Seeni Aatha)

STRUCTURE Small tree. Bark pale, and cracked in older plants. **FLOWER** Yellow-green flowers with three petals. **LEAF** Oval leaves, blunt at tips with usually no suggestion of pointed tip. Parallel lateral veins. Smaller leaves than in the Soursop (see opposite). **FRUIT** Light green, often with glaucous covering. Fruit distinctive, with scales or bumps. **HABITAT & DISTRIBUTION** Tropical American tree widely planted in home gardens for its fruits. **ETYMOLOGY** *Squamosa* is a Latin reference to being covered in scales.

Indian Willow ■ *Polyalthia longifolia*
(S: Ovila)

STRUCTURE Tall tree with distinctive long leaves and drooping branches. Much branched in native form found in the wild. Leaf shape and drooping habit reminiscent of the Weeping Willow of Europe, but this is in a different family (the Salicaceae, or willows and poplars).

FLOWER Greenish-yellow flowers in umbel-like racemes. Flowers look like six-rayed stars. **LEAF** Lanceolate leaves arranged in two ranks, alternately on slender stems. Dark green above and paler underneath. Wavy edges. **FRUIT** Smooth, ellipsoid fruits ripen from yellow-white, to reddish and purplish, and finally black. They are taken by bats. **HABITAT & DISTRIBUTION** Riverine tree in the dry zone. Horticultural variety widely planted in avenues in cities. **ETYMOLOGY** *Polyalthia* is derived from the Greek *polus* (much), and *althein* (to heal), a reference to the plant's medicinal value. *Longifolia* is a reference to the long leaf.

LAURACEAE (BAY-LAURELS)

This family has a worldwide distribution, with about 45 genera and around 2,850 species. With the exception of one genus of parasitic twiners, *Cassytha*, they are aromatic trees and shrubs. Their leaves are typically simple, but can vary and include lobed leaves. The radially symmetric flowers (actinomorphic) are usually in axillary panicles, and can be unisexual or bisexual. The flower parts are in sets of three (trimerous). There can be many stamens, sometimes in four whorls, with the innermost whorl being sterile (staminodes). One whorl typically has glands at the base. The ovary is superior in almost all the species with a single carpel, and the fruit is usually a berry or drupe with a single seed. The flower base and stalk are often enlarged to form a leathery or woody cup that partially embraces the fruit.

Bay-laurels include plants such as the Avocado, which has become increasingly popular in the last decade in cities such as London, resulting in huge commercial plantations in warm climates. Lauraceae genera include *Actinodaphne*, *Litsea* and *Neolitsea*, which contribute to Sri Lanka's forest flora, including some prominent trees in the cloud forests. The genus *Cinnamomum* includes the well-known Ceylon Cinnamon (*Cinnamomum verum*, sometimes known as *C. zeylanicum*). The family name is derived from the Latin *laurus* (bay tree).

Avocado ■ *Persea americana*
(S: Ali Gata Pera)

STRUCTURE Medium-sized tree, typically branching above 5–6m and forming dense canopy. Greyish bark fissured and cracked. **FLOWER** Small, greenish-yellow flowers in axillary panicles. **LEAF** Simple, elliptic leaves have short, pointed tips, at times scarcely noticeable. Midvein and lateral veins well marked and stand out especially on undersides. Leaves alternately arranged and clustered at ends of branches. **FRUIT** Large, oval fruit with green skin. Creamy, soft pulp enclosed in hard seed. Originally often eaten locally with sugar and milk as a desert, but until recently not widely eaten in the West. **HABITAT & DISTRIBUTION** Tree from Central and South America that is widely planted in home gardens. In some parts of Asia, large plantations are grown to meet the demand for the fruit Sri Lankan rainforests do have a native tree in this genus, the Ululu *P. macrantha*, which has long, unequal leaves and white fruits. **ETYMOLOGY** *Persea* is the Greek name for an Egyptian tree with sweet fruits. *Americana* refers to the tree's geographical origin in the Americas.

> ## PANDANACEAE (SCREWPINES)
> The screwpines comprise shrubs, climbers and trees found in the tropics and subtropics, growing from sea level to 3,000m. There are about 900 species in three genera. Some species in the genera *Pandanus* and *Frecynetia* are used locally as food. The trees grow in thickets or singly. Their stems are ringed with leaf scars. Some shrub and tree species have adventitious roots that help prop them up. The leaves are arranged vertically in three (tristichous) or four rows, one above the other. The unisexual flowers are naked. Armed with teeth on the margins and sometimes on the leaf blade, the leaves can be formidable.

Marsh Screwpine ■ *Pandanus kaida*
(S: Weta Keiya)

STRUCTURE Many-branched shrub festooned with thorny leaves. Some plants so densely covered with leaves that branches within are not visible. **FLOWER** *Pandanus* plants are dioecious. Separate male and female plants contain flowers of a single sex. Male flowers grouped together in what looks like a furry catkin. Female flowers also clustered together, on what can appear like a fat spike or long, slender cone of flowers. **LEAF** Long, linear leaves

with row of spines on edges. Leaves often bent at nearly right angles, and arranged spirally at tops of stems, giving rise to the name screwpine. **FRUIT** An aggregated fruit or syncarp made up of many single fruits (drupes) joined together (note that technically, a fruit is an individual seed enclosed in a fleshy covering, known as a drupe). **HABITAT & DISTRIBUTION** Close to water in swampy habitats and paddy fields, and on stream banks. Mainly in the lowlands. **ETYMOLOGY** *Pandanus* is a Latinized form of a Malay local name.

Sea Screwpine ■ *Pandanus odoratissimus*
(S: Mudu Keyiya)

STRUCTURE Large bush seen by the seaside. Prop roots prominent, sometimes looking like a tangled mass. Trunk more likely to spread out into multiple branches than in the Marsh Screwpine (see opposite). **FLOWER** See previous species. Male flowers very fragrant and decay soon after maturing. In India, flowers are used as temple offerings and also worn by humans. **LEAF** Midrib of leaf on underside contains spines, which are also found on leaf edges. **FRUIT** Aggregate fruit (see Marsh Screwpine), turning from green to orange-red as it ripens. **HABITAT & DISTRIBUTION** Seashore plant, dispersed by the tide. **ETYMOLOGY** *Pandanus* is a Latinized form of a Malay local name. *Odoratissimus* refers to the sweetly scented flowers.

Arecaceae (Palms)

What was previously the Palmaceae has been reorganized extensively, using modern DNA analyses, into five subfamilies: the Calamideae (e.g. *Calamus*), Nypodeae (e.g. *Nypa*), Coryphoideae (e.g. *Borassus*), Ceroxyloideae (e.g. *Ceroxylon*) and Arecoideae (e.g. *Areca*). There are about 2,400 species in close to 200 genera. Palms have woody stems often with clear leaf scars. They may be single- or multi-stemmed, though the best-known commercially grown palms are single-stemmed. The leaves are arranged at the top of the stem. The petioles have sheaths, in some species with a spine above the sheath. The leaflet blades are palmately or pinnately veined, and leaflets are V shaped in cross-section. In some species the terminal leaflets are replaced by spines that are used for climbing. The flowers are unisexual or bisexual, typically with three fused or free sepals, and there are as many free or fused petals as sepals. There are typically six stamens, but they can be numerous. The fruits can be single seeded (as in coconuts) or have up to 10 seeds. They typically have a fibrous outer covering (mesocarp) and a thin, woody shell (endocarp) around the seed.

Palms are pollinated by wind and insects, and have a pantropical distribution. Despite their association with blue skies and sandy beaches, they reach their greatest diversification in montane rainforests. The family is also found in subtropical and temperate latitudes in the northern and southern hemispheres. It is one of the most commercially useful plant families, and the recent trend for the consumption of coconut water has further reinforced the palms as an important consumable. The family sets a few angiosperm (flowering plant) records. The Coco-de-mer's 30kg fruits are the largest angiosperm fruits. *Calamus manan* has the longest stems – its climbing stems can reach 200m in length. *Corypha umbraculifera* carries the largest inflorescence of any plant, up to 7.5m long and bearing about 10 million flowers.

Betel Nut ■ *Areca catechu*
(S: Puwak; T: Kamukai)

STRUCTURE Thin, tall palm with circular trunk and distinct rings on bark showing leaf scars. Bark grey or greyish-white. **FLOWER** Yellowish flowers on inflorescence that arises between leaves. Flowers are in a panicle. **LEAF** Cluster of leaves at top of growing stem.

Numerous leaflets oppositely (pinnately) arranged along midrib. Leaflets have several distinct parallel veins. **FRUIT** Egg-shaped orange fruit. Seed is chewed with leaves or flowers of *Piper betle* (family Piperaceae), lime and other ingredients. The alkaloid arecaine in the seed is a mild narcotic. **HABITAT & DISTRIBUTION** Wet lowlands and the intermediate zone. Popular plant in home gardens. **ETYMOLOGY** *Areca* is derived from the local name in Malabar, India. *Catechu* is a Malayan name.

Palmyrah Palm ■ *Borassus flabellifer*
(S: Tal; T: Panai)

STRUCTURE Tall, fan-leaved palm with dark trunk. **FLOWER** Male flowers clustered in compact, branching spikes. Female flowers on separate female trees; larger flowers in scattered, sparingly branched clusters. Male trees are tapped for toddy. **LEAF** Large, fan-shaped leaves that are palmately veined. Leaf stalks shorter than leaves; a point of difference with palms in the genus *Corypha*. Fruits and flowers also differ. Dried leaves are used in thatching and for making boundary walls. **FRUIT** Dark, spherical fruit. Much smaller than coconut. Toddy is an alcoholic drink made from fermented fruits. **HABITAT & DISTRIBUTION** The trees are a distinctive part of the landscape in Mannar, some parts of the east coast and the Northern Peninsula. They thrive on dry sandy soils, especially close to coasts, in an environment that is hostile to other trees. **ETYMOLOGY** *Borassus* is derived from Greek, and is a reference to a part of a palm tree. *Flabellifer* means to bear fan-shaped leaves.

Fishtail Palm ■ *Caryota urens*
(S: Kithul; T: Tipplilipan)

STRUCTURE Slender, tall palm. **FLOWER** Single-sex flowers hang in enormous clusters likened to horse tails. **LEAF** Leaves twice pinnate (bipinnate). Opposite pairs of leaflets divided again into pairs with irregular lobes, giving the impression of a fishtail, hence the common name. Unlike in other palms, leaves are not all crowded at top of stem, and many arise a distance away from the top. **FRUIT** Small, round red berries hang together in pendulous strings. **HABITAT & DISTRIBUTION** Widely planted in home gardens in wet zone. Frequently visited by the endemic Ceylon Hanging-parrot. **ETYMOLOGY** *Caryota* comes from the Greek *karuotos* (nut-like). *Urens* comes from Latin, meaning burning or stinging.

Coconut ■ *Cocos nucifera*
(S: Pol; T: Tennai)

STRUCTURE Tall, slender palm tree with leaves at the top. One of the most familiar and recognizable trees in the world. Bark greyish-white with prominent rings marking positions of fallen leaves. Every part of the tree has a use. **FLOWER** Cream flowers in branching raceme enclosed in enveloping sheath (spathe). **LEAF** Pinnate leaves have leaflets that are long and narrow with pointed tips. Midveins of leaflets, known as *eakles*, used in *eakle* brooms to sweep gardens. **FRUIT** Large, rounded fruit with thick, fibrous husk enclosing seed that contains coconut water, a refreshing drink. White kernel utilized to prepare 'coconut milk', used widely in Asian cuisine. **HABITAT & DISTRIBUTION** Planted all around the coast where there is adequate rainfall to support commercial plantations. Also popular tree in home gardens. **ETYMOLOGY** *Coco* means head in Spanish and is a reference to the seed with its three holes resembling a head. The name was given in the 16th century.

Wild Date Palm ■ *Phoenix pusilla*
(S: Indi; T: Inchu)

STRUCTURE Slow-growing, small tree, often found in small thickets in dry zone.
FLOWER Male and female flowers occur on separate trees. Male flowers white and clustered around fleshy axis (spadix), which has short stem (pedicel). Creamy-white

female flowers clustered on spikes emanating from long stalk. **LEAF** Pinnate leaves with long, pointed leaflets. Lower leaflets form spines. **FRUIT** Egg-shaped, bright red to purple fruits. The date palm *P. dactylifera*

has been cultivated in the Middle East and Africa for several centuries as an easily transportable food with high sugar. It is not found in the wild, and the plant known today may be a hybrid species. **HABITAT & DISTRIBUTION** Plant of the dry lowlands, common in the north. **ETYMOLOGY** *Phoenix* is derived from the Greek word for purple, a reference to the fruits.

Musaceae (Bananas)

This is a well-known family thanks to the ubiquitous banana. It has a wide natural distribution in the tropics, from Africa, across Asia, to New Guinea, Queensland and Melanesia. There are three genera with about 80 species, a majority of which (about 70) are in the genus *Musa*. The family comprises giant herbs and strictly speaking does not belong in a book about trees, which are plants with woody tissue to support their structures, but it is included because of the Banana's tree-like form.

The Banana is a curious plant in that a stem dies after flowering and fruiting (monocarpic, meaning single fruit). However, new stems spring up from underground rhizomes and the plant stays alive for multiple seasons of fruit and flower. The alternate leaves are enormous, and their leaf sheaths form a false stem at the base. The flowers grow in a terminal thyrse. In the usual formation, the inflorescence contains both male and female flowers, with the females at the base and the males near the apex. The flowers have bilateral symmetry with six petals. However, five of the petals are fused to form a five-lobed sheath. Banana 'trees' have become popular in European parks and gardens (with cold-tolerant horticultural varieties) because of their showy foliage and dramatic inflorescences.

Banana ■ *Musa acuminata*

STRUCTURE Large, thick trunk, long, wide leaves and the familiar fruits. **FLOWER** Cone-like flowering, with flowers sheathed in a purplish leaf-like envelope. Interesting flower with upper part male and lower part functionally female, though lower part is bisexual. Flowers after about 15 months. Female flowers grow into finger-like fruits in tiers while male flowers are still present. **LEAF** Large leaves long and blunt at ends. Stout midrib has many parallel veins running perpendicular to midrib. Old leaves often ragged and torn. Leaves paler underneath. They are often used in Sri Lanka as a biodegradable wrapper for food. *Lamprais*, a rice and curry dish with Dutch influence, is wrapped in banana leaves. **FRUIT** Long, fleshy fruit with seeds embedded inside. Seeds absent in some cultivated varieties. One of the most familiar fresh fruits in the world. Fruit of some varieties has to be cooked for consumption as a vegetable. **HABITAT & DISTRIBUTION** Grown widely in home gardens, and commercially in large plantations.

Wild plant believed to have originated in America, though some authorities maintain that a native species grows in the eastern Himalayas and other parts of Asia. **ETYMOLOGY** Genus name *Musa* honours Antonius Musa, physician to Octavius Augustus, the first Emperor of Rome.

DILLENIACEAE (DILLENIAS)

The dillenia family is found in tropical and subtropical regions, and comprises about 450 species in 11 genera, which are widely distributed in Australia. The genus *Hibbertia*, popular as a garden plant, ranges from Madagascar to Australia. Most species are woody, growing as shrubs, trees and lianas. Some have edible fruits. Their flowers are showy, radially symmetric and have 3–5 petals with numerous stamens. Many species use buzz pollination. The leaves are simple and alternate, often with prominent veins and dentate margins. The fruit is often enclosed by fleshy sepals. *Dillenia* tree timber is used in boat-building.

Blunt-leaved Dillenia ▪ *Dillenia retusa*
(S: Godapara)

STRUCTURE Small tree with thick canopy. Trunk brownish-red. **FLOWER** Large white flowers on long stalk. **LEAF** Large leaves, blunt and rounded at tips. Leaves fatter in terminal half (obovate), and very strongly veined, with several parallel veins running from midrib to leaf edge. Leaf edge simple. Leaf stalk swollen at base, narrowing to where it joins leaf blade, and channelled above.

FRUIT Round, slightly compressed fruit orange on ripening. Finely hairy. Seeds enclosed within pulp. **HABITAT & DISTRIBUTION** Common in wet zone in disturbed areas. **ETYMOLOGY** The Swedish taxonomist Linnaeus named the genus *Dillenia* in honour of J. J. Dillenius, professor of botany at Oxford and a contemporary of his. *Retusa* is Latin for blunt tipped.

Three-angled Dillenia ■ *Dillenia triquetra*
(S: Diyapara)

STRUCTURE Small tree superficially similar to the Blunt-leaved Dillenia (see opposite). Leaf shapes different. Smooth, brownish bark marked with leaf scars. **FLOWER** Flowers borne on racemes. Prominent white flowers have yellow centres with stamens. Fleshy sepals

persistent on fruits. **LEAF** More rounded than in Blunt-leaved, elliptic and almost circular. Leaves have pronounced wavy margin. Leaf stalk channelled above and swollen at base. Young leaves have prominent horseshoe-shaped, fleshy cushion on upper leaf stalk. **FRUIT** Small, round fruit enclosed by enlarged sepals. **HABITAT & DISTRIBUTION** Widely distributed in wet zone in disturbed areas and where secondary forests have become established. **ETYMOLOGY** *Triquetra* refers to being three angled or three sided.

Kekiriwara ■ *Schumacheria castaneifolia*
(S: Kekiriwara)

STRUCTURE Takes the form of large, straggling shrub or small tree. Frequently encountered small tree along former logging road used by nature tourists in Sinharaja from the Kudawa entrance. **FLOWER** Small, numerous yellow flowers in panicles, usually arising from leaf axils. The five sepals are stiff, and three are longer than the other two. Typically four petals, which tend to shed early, and 25–30 stamens. **LEAF** Lance-shaped, blunt-tipped leaves. Other members of this genus in Sri Lanka have leaves that are more oval or elliptic in shape, but given the variability of leaf shape there is some doubt if the others are valid species. Parallel leaf venation shows clearly. **FRUIT** Fruit splits to distribute seed that is covered with a membrane (aril). **HABITAT & DISTRIBUTION** Endemic plant common in wet-zone forest patches. Often seen on forest edges of disturbed or secondary forests. Absent where forests have been cleared.

CELASTRACEAE (SPINDLE TREES)

This family comprises nearly 100 genera and around 1,200 species. There have been a lot of taxonomic changes with regard to which plants are members of the family based on genetic analyses. The family is divided into two major subfamilies, the Parnassioidea and Celastroidea. The latter includes all but 70 or so of the species.

The spindle tree family contains plants with a wide range of growth forms, including annual and perennial herbs, climbers, epiphytes, shrubs and trees. The plants can be unisexual or bisexual, and the stems are thorny in some species. The form and arrangement of the leaves vary widely across the family. Some species have hollow structures (domatia) for ants. The arrangements of flowers also vary, from arising on the stems (cauliflory), to occurring in axillary racemes, thyrses and so on. Most species have radially symmetrical flowers, although a few have flowers that are bilaterally symmetrical. The flowers have 4–5 sepals, and petals may be absent or number 3–7. The stamens number two, three, four or five, and are fixed between the petals. The fruits can take many forms, and some species have edible seeds or fruits while others are poisonous.

Egg-fruit ■ *Pleurostylia opposita*
(S: Panakka; T: Chiru-piyari)

STRUCTURE Attractive medium-sized tree with broad, spreading green canopy. Bark greyish with small reticulations. **FLOWER** Small, greenish flowers in axillary cymes. Calyx five lobed. Five petals. Fleshy flower disc has five stamens with their bases attached beneath it. Ovary shaped like a flask. **LEAF** Stiff, dark green leaves arranged in opposite pairs. **FRUIT** Small, egg-shaped, prominently white fruit. **HABITAT & DISTRIBUTION** Found in dry lowlands, and neither rare nor especially common. Seems hardy enough to establish itself on sandy soils. The only member of its genus in Sri Lanka, the species occurs from Madagascar, India and Southeast Asia, to Australia and New Caledonia. **ETYMOLOGY** *Opposita* is probably a reference to the opposite leaves.

OXALIDACEAE (WOOD SORRELS)

This family of about 800 species is found mainly in tropical and subtropical regions. The plants occur as annual or perennial herbs, climbers, shrubs and small trees. Despite the large number of species, they are all placed in 5–6 genera (depending on the author). The family includes the edible Star Fruit, but is dominated by weedy species in the genus *Oxalis*. Some species grow in clumps spreading on stolons. The plants have five sepals and five petals, but the inflorescences are variable from species to species. The sexual arrangement also varies, with some species being unisexual and others bisexual.

Bilin ▪ *Averrhoa bilimbi*
(S: Bilin)

STRUCTURE Much-branched small tree with feathery leaves spirally arranged at ends of branches. Bare trunk has fruits growing from it (this is cauliflory, considered a primitive feature of plants). **FLOWER** Small, symmetrical red flowers with five red sepals and petals, and 10 stamens. Ovary has five styles and five cavities. Flowers on trunk and large branches. **LEAF** Several (eight plus) pairs of pinnate, oblong leaflets with short stalks, with terminal leaflet. **FRUIT** Green, cylindrical, soft and juicy fruit. Sour to the taste but ripe fruits are eaten uncooked in Sri Lanka as a snack with salt and chilli. Also used in pickles. **HABITAT & DISTRIBUTION** Probably originated in the Indonesian Moluccas. Widely grown in the tropics. In Sri Lanka found in home gardens in the lowlands, and grows especially well in the wet zone. **ETYMOLOGY** Genus name honours the Arab philosopher Averhoes (AD 1126–1198).

Star Fruit ■ *Averrhoa carambola*
(S: Kamaranga, Carambola)

STRUCTURE Small tree that branches profusely. **FLOWER** Fleshy flowers reddish-purple in centre and paler at edges. Flowers form on trunk (cauliflorous), or in axils of young branches. **LEAF** Short-stalked, oblong leaflets arranged pinnately in pairs with terminal leaflet. Leaflets pointed, with terminal leaflet the biggest. **FRUIT** Fruit star shaped in cross-section, hence the common name of Star Fruit. Ripens to yellow, and can be sour or sweet. **HABITAT & DISTRIBUTION** Native to Southeast Asia and Indian subcontinent, and introduced to Sri Lanka. Found in home gardens in the lowlands. **ETYMOLOGY** Genus name is the Spanish name for the plant.

RHIZOPHORACEAE (MANGROVES)
The mangrove family comprises evergreen trees and shrubs found worldwide on tropical coastlines with warm, shallow seas. In Sri Lanka mangroves are seen where sea water mixes with fresh water to form estuarine habitats. Many species have stilt roots that allow them to obtain oxygen when the roots are flooded. Their leathery leaves are usually simple. Interpetiolar stipules sheath the terminal buds and fall off as a leaf grows. The flowers are in axillary racemes, or arise in a cluster from the same point (fascicle). They can have a cup-shaped extension of the basal part holding the sepals, petals and stamens (hypanthium). The flowers are bisexual and radially symmetric (actinomorphic). The number of sepals varies at 3–26, and they are free or fused at the base. There are as many petals as sepals. The stamens are often a twice times multiple of the petals, and the filaments are fixed around the base of a nectary disk. The ovary is superior to inferior with 2–5 carpels, and the leathery or pulpy fruit can be a berry, drupe or non-splitting (indehiscent) capsule.

Long-stalked Kadol ■ *Rhizophora mucronata*
(S: Kadol)

STRUCTURE Small tree found in mangroves with prop roots. Closely related *R. apiculata* has stalkless (sessile) flowers, whereas *R. mucronata* flowers have long stalks. Large number of visible prop roots. **FLOWER** White flowers with yellow sepals, in fours, in leaf axils. **LEAF** Broad leaves, elliptic to oval, and thick. Dark glandular dots visible on undersides – a trait of plants in this genus. **FRUIT** Long pod begins to germinate before fruit has ripened, a feature known as viviparity (meaning the bearing of live young). This is an adaptation to head-start the young plants that drift off on the tide until they become embedded in mud in a suitable place to start growing. **HABITAT & DISTRIBUTION** The dominant *Rhizophora* species from Negombo and the Kalpitiya Peninsula northwards. Between Colombo and Matara, the *Rhizophora* species that is dominant is *R. apiculata*. Further south in the dry-zone mangroves around Rekawa, *R. mucronata* becomes dominant. Although both species can be found together, there is a clear habitat preference for mangroves in dry coastal areas by *mucronata*, and for mangroves in wet humid regions by *apiculata*. **ETYMOLGY** Mucronata is a reference to the mucronate tip, a slender, pointed tip in an otherwise oval leaf.

CLUSIACEAE (MANGOSTEENS)
Largely a pantropical family, the mangosteens form an economically important group of plants that grow as herbs (annual or perennial), shrubs and trees. They provide fruits, latex, essential oils, drugs, pigments and resins. The mangosteens have glands or canals in most parts of the plant, from which secretions take place. Their leaves are simple, usually without stipules and usually opposite, but sometimes alternate or whorled. The radially symmetric flowers can be unisexual or bisexual in different species. The flowers are solitary or borne in a cyme at the end of a shoot or in axils. The fruits can be fleshy or dry berries, or drupes. *Callophyllum* and *Mesua* species are important for their timber, while *Garcinia* species have edible fruits, including the familiar mangosteen.

Domba ■ *Calophyllum inophyllum*
(S: Domba)

STRUCTURE Short tree that can grow to 20m tall. Much branched, with dense green canopy. Bark greyish-brown and cracked. **FLOWER** White flowers in erect racemes. Many stamens crowded around a style. **LEAF** Elliptic, thick leaf with strong midvein. Clear lateral veins. Leaves large; bigger than a person's hand. **FRUIT** Green spherical fruits suspended on long stalks from leaf axils. Fruit bats are an important agent of dispersal. **HABITAT & DISTRIBUTION** Typically a tree of the seaside, but often planted inland. **ETYMOLOGY** *Calophyllum* derives from the Greek *kalos* (beautiful) and *phullon* (leaf). *Inophyllum* is a Greek reference to a leaf with pronounced veins.

Edible Goraka ■ *Garcinia quaesita*
(S: Kana Goraka, Hinda Goraka)

STRUCTURE Much-branched small tree forming dense canopy. White latex helps to distinguish it from other small forest trees that have opposite leaves, in the *Eugenia*, *Memecylon* and Rubiaceae. **FLOWER** White flowers on axils of upper leaves; males clustered, females solitary. In *Garcinia* species, stigma rests on ovary and not on end of a style. **LEAF** Oval leaves, lightly scalloped on edges. Young leaves red. **FRUIT** Reddish berry with multiple grooves. Pulp is acidic. **HABITAT & DISTRIBUTION** Occurs naturally in the understorey of intermediate forests and rainforests. Widely planted in home gardens. **ETYMOLOGY** Genus name honours Laurence Garcin (1683–1751), a Swiss botanist who collected plants in India.

Ironwood Tree ■ *Mesua ferrea*
(S: Na)

STRUCTURE Small tree with slightly drooping branches and somewhat untidy appearance. **FLOWER** Large white flowers with yellow centres and many stamens. White petals crinkled at edges. **LEAF** Oblong to lanceolate leaves with pointed tips. Midrib prominent, lateral veins visible. New leaves flushed red. **FRUIT** Spherical brown capsule enclosed by enlarged sepals and bracts. **HABITAT & DISTRIBUTION** Occurs naturally in rainforest subcanopy near water. Widely planted as an ornamental to line avenues. **ETYMOLOGY** Genus name honours J. Mesue (AD 777–857), a polymath better known as John of Damascus.

SALICACEAE (WILLOWS)

This family of trees and shrubs includes a few species that are nearly herbaceous if not for a woody base. There are about 1,220 species in 56 genera. Their simple leaves are arranged alternately, sometimes in two ranks (distichous). In some species the veins end near the margin with a dark spherical gland. The leaf stalks (petioles) also bear glands. The flowers can aggregate as racemes, spikes, catkins or clusters at one point (fascicles), and each flower has a bract clasping it from underneath (subtending). The flowers can be unisexual or bisexual, and the arrangement can vary with one or both occurring on the same plant. The sepals can vary from none to 15, but typically number 3–8. They can be free or basally fused. At the base of the flower is a nectary disk that is often lobed. The ovary is superior, with 1–10 carpels. The fruits are berries, drupes or capsules with three valves. Some of the well-known species in the genera *Populus* and *Salix* have silky hairs on the seeds that aid with wind dispersal. The family includes species that were previously in the Flacourticaceae family.

Governor's Plum ▪ *Flacourtia indica*
(S: Uguressa)

STRUCTURE Typically a bushy shrub, but can also grow as a small tree. Rough bark with raised flakes and interrupted occasionally with spines. Branches have spines. **FLOWER** Very small, greenish-yellow flowers in clusters within leaves. Petals absent, and sepals greenish-yellow. Many crowded yellow stamens in male flowers. Female flowers have 6–7 styles. Flowers are unisexual, and male and female flowers occur on separate trees (dioecious). Tiny flowers are easily overlooked. **LEAF** Elliptic to obovate. Cluster of leaves on short stalks at nodes on branchlets. Leaves blunt tipped, and at times slightly notched (emarginate). **FRUIT** Smooth, purplish berry with several seeds. **HABITAT & DISTRIBUTION** In lowlands in rainforests and monsoon forests as an understorey plant. Planted as a thorny hedge in India. **ETYMOLOGY** Genus name honours E. de Flacourt (1607–1660), general director of the East India French Company and a governor of Madagascar. *Indica* refers to India.

Lovi ▪ *Flacourtia inermis*
(S: Lovi)

STRUCTURE Typically a small tree but can grow to up to 15m in height. Bark grey-brown. **FLOWER** Tiny green-yellow flowers in clusters. **LEAF** Elliptical leaves scalloped or toothed on edges. Young leaves flushed red. **FRUIT** Spherical green berries in bunches, turning red when ripe. Pulp is acidic. **HABITAT & DISTRIBUTION** Introduced tree from Malesian region (Malaysia, Indonesia and so on). Widely cultivated in home gardens in the lowlands for its edible fruits.

EUPHORBIACEAE (SPURGES)

This widespread family is absent only in the Arctic and Antarctic regions. It includes around 6,250 species in just over 200 genera. The white latex that occurs in some species is a characteristic of the family, and is very obvious and well known in the rubber trees (*Hevea* species) that are planted on an industrial scale for natural rubber. However, many species in the family lack latex. The family comprises plants that look very different – so much so that from external appearances, to the untrained eye, it is hard to imagine that some of the species are in the same family.

The plants' leaves are generally simple, but compound leaves (typically palmate) occur in a few species. Stipules are present. The flowers (mainly green, yellow or white) are usually tiny and difficult to examine without a hand lens, and they are male or female. There are 3–5 sepals and as many petals, though more often petals are absent. The number of stamens is variable. If the superior ovary has a single style, it will be divided into 2–3 arms, with each arm forked into two. Alternatively, there may be 2–3 styles in the flower, each forked once. The fruit is typically a capsule with three shoulders, or lobes, and three cavities, with each cavity containing 1–2 seeds. In some species, the fruit is a small, pulpy berry. The arrangement of the flowers can be highly variable, ranging from racemes, panicles and spikes, to heads and catkins. In some species the male and female flowers are on different trees (unisexual), so that there are separate male and female trees. In others, the male and female flowers are in the same inflorescence but arranged in a standard way in relation to each other.

Candlenut ■ *Aleurites moluccana*
(S: Tel Kekuna)

STRUCTURE Medium-sized evergreen tree with brownish-grey bark. **FLOWER** Large numbers of small white flowers form on ends of branches. Separate male and female flowers on the same tree (monoecious). Female flowers have three styles, and male flowers have many stamens. **LEAF** Leaves can take one of two shapes (or intermediate variations). They are either egg shaped, or with three lobes and veins radiating from where leaf stalk joins leaf blade (palmate veins). Leaves large. **FRUIT** Fleshy berry with green husk. Inside is a seed with an edible kernel. Tung oil is extracted from the seed, and strings of seeds can be used as candles. **HABITAT & DISTRIBUTION** Native to the Malay Archipelago and introduced to other tropical countries. **ETYMOLOGY** *Aleurites* means floury in Greek, and *moluccana* refers to the Moluccas.

Hedge Boxwood ■ *Drypetes sepiaria*
(S: Weera; T: Weerai)

STRUCTURE Medium-sized tree with fluted trunk. Many trees are multi-stemmed, lending the appearance of a tree-like bush. Often found in large groves where it is the only tree. Bark greyish and smooth. **FLOWER** Flowers yellowish-white. Male flowers in axillary clusters or racemes. Male flowers have 4–23 stamens; female flowers are in ones or twos. **LEAF** Dark green leaves have simple margins and rounded tips and bases. Sides are somewhat parallel. Leaves stiff (coriaceous). **FRUIT** Round berries turn bright red on ripening. An important food source for birds and mammals. Fruits single seeded. **HABITAT & DISTRIBUTION** Native to southern India and Sri Lanka. Common and important tree in monsoon forests of the dry lowlands. Familiar tree to those visiting national parks such as Yala and Wilpattu.

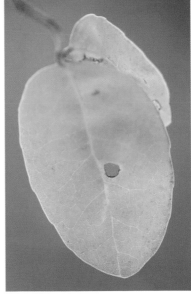

Daluk ■ *Euphorbia antiquorum*
(S: Daluk; T: Chatura kalli)

STRUCTURE Grows as a small tree or bush. Fleshy green branches result in this tree being confused with a cactus, which it closely resembles. Fleshy branches have a milky juice. Branches jointed, with undulating ridges with spines. Branches have 3 or 5 angles.
FLOWER Pale yellow flowering heads at ends of branches. Single female flower surrounded by male flowers in a cup-shaped receptacle (involucre). Male flowers have one stamen

each. Style in female flower is free. **LEAF** Tiny leaves fall early and the plant is often without leaves. **FRUIT** Green capsule with three lobes. Somewhat triangular in cross-section. **HABITAT & DISTRIBUTION** Native plant growing wild in the dry lowlands. Also grown as a hedge plant or an ornamental. **ETYMOLOGY** Pliny stated that *Euphorbia* honoured a physician of King Juba of Mauritania. *Antiquorum* in Latin means 'of the ancient writers'.

Kenda ■ *Macaranga peltata*
(S: Kenda; T: Vatta kani)

STRUCTURE Old individuals can grow to about 30m in height to be medium-sized trees. Most trees seen on roadsides tend to be small and almost shrub-like. Bark has horizontal bands of lenticels. **FLOWER** Greenish flowers are in panicles. Separate male and female flower inflorescences. Male flowers on inflorescences around 10cm long and flowers have no stalk. Female inflorescences around 3cm long and flowers have tiny stalks, about 2mm in length. **LEAF** Leaf a rounded triangle with sharp point on tip. Leaf blade attached below to long leaf stalk. Leaves have nectaries near to where stalk joins leaf blade. On close inspection, ants may be seen feeding on the nectar (the ants protect the tree from herbivores). Leaves used as a wrapper when cooking a local sweet named *Halapa*. **FRUIT** Small fruit tiny and round, with one side having the remains of a stigma (persistent stigma). **HABITAT & DISTRIBUTION** Found in southern India and Sri Lanka. Very common in the wet lowlands, below 1,000m. **ETYMOLOGY** *Macaranga* is derived from a local name in Madagascar. *Peltata* is a reference to the leaf petiole being attached towards the centre of the leaf blade and not at the edge of the blade, a feature common to all of the macarangas.

Indian Gooseberry or Amla ■ *Phyllanthus emblica*
(S: Nelli; T: Topinelli)

STRUCTURE Small to medium-sized tree with an overall feathery appearance. Seasonally sheds leaves (deciduous). Bark on branches smooth and greenish-grey. Bark on trunks of old trees fissured. If bark is cut it reveals a deep red colour. **FLOWER** Male and female flowers on the same branch. Male flowers have short stalks and are numerous. Female flowers fewer in number, lack stalks and look as though they are pressed close against branches. **LEAF** Leaves arranged in opposite pairs along branches, so a branch can resemble a doubly pinnate leaf. Each leaf has small leaflets that are opposite each other. Leaves and leaflets within leaves closely arranged in a single plane, accentuating a

feathery look. **FRUIT** Fleshy pulp surrounds hard nut containing six seeds. Fruit has six shallow lobes. Green fruits turn yellow or pink when ripe. Fruit acidic. **HABITAT & DISTRIBUTION** Native to India, Sri Lanka and Malaya. Also grown in home gardens. Fruits are used to make chutneys, and various parts of the plant are used medicinally. **ETYMOLOGY** *Phyllanthus* is derived from the Greek *phullon* (leaf) and *anthos* (flower). Emblica is derived from a local Indian name.

LINACEAE (FLAXES)
The flaxes are a diverse family growing as shrubs and trees. They comprise about 250 species in 14 genera. The majority are herbs growing in the northern temperate zone, and many are slender and hairless. The Flax is the source of linseed oil. It is also used for its fibres for making good-quality paper. The plants are smooth, with alternate or opposite leaves without petioles. Radially symmetrical flowers are borne in inflorescences (spikes or cymes). There are five sepals and five petals, and the latter fall off early.

Moderakanniya ■ *Hugonia mystax*
(S: Moderakanniya)

STRUCTURE Untidy shrub that can grow to 7m in height. Branches climb using hooks.
FLOWER In this species flowers are 2.5–4cm, about twice the size of those of *H. ferruginea*. **LEAF** Differs from *H. ferruginea* by its hairless, oval-shaped leaves crowded at ends of twigs. *H. ferruginea* is distinct, with brownish hair and silky undersides to leaves, and leaves are not crowded. In this species, somewhat stiff leaves marked with distinct midrib. Veins reticulated on both surfaces (arched in *H. ferruginea*). Five sepals and five yellow petals. Ten stamens alternately long and short. Five styles longer than stamens and club headed (capitate).
FRUIT Round berry turning from yellow to red as it ripens.
HABITAT & DISTRIBUTION Found in the dry lowlands.
ETYMOLOGY Genus name honours Johann von Hugo (1686–1760), a German physician and naturalist.

FABACEAE (LEGUMES)

The peas, or legumes, are one of the best-known plant families, key to providing food worldwide for humans and second only to the cereals, especially because of their consumption worldwide. There are about 20,000 species in more than 700 genera, occurring as herbs, annual shrubs and trees. Lupins are used in horticulture for their showy flowers, the groundnut (peanut) is an important food plant and robinias are planted widely in cities around the world. Clitorias are common climbers often colonizing rough ground in cities and possessing showy, butterfly-like flowers. Next to the daisies (Asteraceae) and orchids (Orchidaceae), the legumes are the third most species-rich family.

A key characteristic of the legume family is the fruit, which is a one-chambered seed pod; the legume. In addition, the root nodules of legumes contain nitrogen-fixing *Rhizobium* bacteria. Traditionally, the family has been divided into three major subfamilies. They have been grouped this way below as this arrangement is the one most familiar to people. However, six subfamilies are now recognized, the Duparquetioideae, Cercidoideae, Detarioideae, Dialioideae, Faboideae and Caesalpinioideae. Plants that were previously in the Mimosoideae are now within the Caesalpinioideae. Most of the species found in Sri Lanka are in the Faboideae and Caesalpinioideae. However, the introduced Pride of Burma is in the subfamily Detarioideae.

The traditional arrangement is as follows:

Caesalpinioideae (Cassias) This subfamily includes some large tropical trees with showy flowers. One of the petals in the flowers may be different from the rest. Well-known genera in the family include *Bauhinia*, *Cassia*, *Delonix*, *Parkinsonia*, *Peltophorum* and *Tamarindus*. The subfamily is predominantly tropical and warm temperate. It has four tribes, Cercidae, Detarieae, Cassiae and Caesalpinae.

Mimosoideae (Acacias) This subfamily includes the genera *Albizia*, *Dichrostachys*, *Enterolobium*, *Leucanea*, *Prospis* and *Samanea*. Many genera include species that have been introduced into Asia. The genus *Acacia* (now split into further genera) held about 1,450 species. The subfamily is predominantly tropical and warm temperate. The mimosas have twice-compound leaves forming feathery leaflets.

Papilionoideae (Peas) The flowers of this subfamily typically have five petals, with the top one being the 'standard', a pair on each side (the wings) and a pair forming a 'keel' at the bottom partially enclosing typically 10 stamens. Well-known genera in this subfamily include *Crotalaria*, *Dalbergia*, *Lupinus*, *Sesbania* and *Vicia*. Unlike the other subfamilies, the Papilionoideae have major areas of diversity in temperate areas. The subfamily is divided into four tribes, Mimosae, Mimozyganthae, Acaciae and Ingeae. It should not to be confused with the butterfly family Papilionoidea, similarly spelt but lacking the 'e' at the end.

Because of their nitrogen-fixing ability, many legume species have been introduced all over the world into plantations. Species of the genus *Prosopis* have become problematic invasive plants as they can thrive on poor soils and grow into dense thickets, excluding native species.

Ear-pod Wattle ■ *Acacia auriculiformis*

STRUCTURE Medium-sized evergreen tree. Bark greyish with fissures. Spines absent.
FLOWER Large number of fragrant, stalkless flowers (sessile) crowded in whorls along long
spikes. Tiny petals fold over on edges (recurved). They are easily overlooked, and the many
long yellow stamens are conspicuous, imparting an overall impression of golden flower
spikes. **LEAF** Leaves are not true leaves (which are only present when a tree is young), but
'phyllodes' – expanded and flattened leaf stems that have photosynthetic matter – take on
the role of leaves. However, they are leathery and lack the vast number of pores that leaves
have, which can result in loss of water. Some acacias have evolved the use of phyllodes
to reduce water loss in arid conditions. Long, slender phyllodes taper to a point and are
curved, with three prominent parallel veins, and a gland at the base of each phyllode.
FRUIT Pod coiled and ear-like. Green at first and turning brown with maturity. Pods split
on tree and seeds hang out, suspended by orange filaments. **HABITAT & DISTRIBUTION**
Native to New Guinea and Australia. Seen in arid zone where it is planted for shade and as
an ornamental. **ETYMOLOGY** *Auriculiformis* in Latin means ear shaped.

Axe-twister ▪ *Acacia chundra*
(S: Rat-kihri; T: Kodali-murunkai)

STRUCTURE Medium-sized tree. Feathery in appearance. Bark dark brown, furrowed deeply in older trees. Strips of bark peel off. Hook-shaped thorns on elongated bases at leaf bases. Very similar to *A. catechu*, but differentiated by smooth leaves and considerably fewer leaflets and pinnae. **FLOWER** Hundreds of tiny flowers packed into spikes. Petals virtually absent. Dense heads of stamens creamy-white to light yellow, creating flowering spikes of that colour. **LEAF** Eight to 10 pairs of leaflets arranged in opposite pairs on a midrib, with each in turn having about 30 pairs of small pinnae that are oppositely arranged (thrice pinnate). Pinnae strap shaped. **FRUIT** Flat, thin pod with slight constrictions. **HABITAT & DISTRIBUTION** Arid zone of the south-east and north-east. The region around Eluvankulam, between Wilpattu and Mannar, is a good area for this tree. **ETYMOLOGY** Red heartwood very hard and used for making axe handles.

Jumble Bead ■ *Adenanthera pavonina*
(S: Madatiya)

STRUCTURE Tall tree with dark grey bark. **FLOWER** Minute flowers, each with 10 prominent stamens with anthers tipped with glands. Lightly scented, creamy-yellow flowers are in racemes, which arise from bases of leaves found at ends of branches. Each flower has five petals that take on a star shape. Calyx bell shaped. **LEAF** Midrib has opposite lateral leaflets along which elliptic pinnae are arranged alternately in two rows (bipinnate with alternate pinnae). Note unusual arrangement, in that pinnae are alternate and not opposite as in most bipinnate plants. There is a terminal pinna. **FRUIT** Pods split (dehisce) with much twisting of the two strap-like sides that split along their edges. Red seeds often seen in split pods on a tree. Note that the pods are not jointed between the seeds, a trait that distinguishes this genus from others in the family. Seeds collected for decorative use, and dispersed by birds. **HABITAT & DISTRIBUTION** Native to India and Southeast Asia. Introduced tree planted in the lowlands. **ETYMOLOGY** *Adenanthera* is derived from the Greek *aden* (gland) and *anthera* (blooming), a reference to the gland on the anthers. *Pavonina* is a Latin reference to coloured or peacock-like (note that peacocks are in the genus *Pavo*).

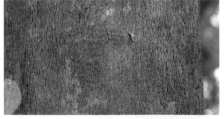

Ceylon Rosewood ■ *Albizia odoratissima*
(S: Sooriya Mara; T: Karu-vakai)

STRUCTURE Tall tree with compact appearance. Canopy light, a feature used to separate it from *A. lebbek.* Bark silvery or greyish. **FLOWER** Fragrant flowers. **LEAF** Opposite pairs of pinnae (pinnate) on a leaflet, with leaflets in turn arranged in opposite pairs (bipinnate).

FRUIT Pod about 20cm long, and often hidden by leaves. **HABITAT & DISTRIBUTION** Native to Sri Lanka, India and Malaysia. This contradicts the perception that all *Albizia* in Sri Lanka are introduced. Occurs in dry areas from the lowlands to middle elevations. Often planted on tea estates. **ETYMOLOGY** Genus name honours the 18th-century Italian botanist F. del Albizzi. *Odoratissima* refers to the fragrant wood.

Pride of Burma ■ *Amherstia nobilis*

STRUCTURE Medium-sized to tall tree with straight trunk and heavy branches. Bark smooth and greyish. In Sri Lanka's wet zone, bark often heavily mottled with algae and lichens. Young leaves hang limply and are vividly coloured in pinks and reds, contrasting with dark green of older leaves. **FLOWER** Beautiful multicoloured, mixed red-and-yellow flowers arranged in racemes that hang (pendulous) from long stalk, and have been likened to a candelabrum. Four petal-like sepals and five petals, with three being large and unequal. Ten stamens with five longer than others. **LEAF** Leaves are pinnate with leaflets set in opposite pairs, and lacking a terminal leaflet (paripinnate). Leaflets pointed. Leaves can be quite long, extending up to 50cm. **FRUIT** Typical bean pod, about 6cm long. Turns bright crimson when ripe. **HABITAT & DISTRIBUTION** Native to Myanmar and introduced into parks and gardens. Grows well in wet zone. **ETYMOLOGY** Genus name honours Lady Sarah Amherst, a 19th-century artist and collector in India. *Nobilis* is derived from the Latin for stately or noble.

Bauhinia ■ *Bauhinia racemosa*
(S: Maila; T: Atti)

STRUCTURE Grows as a small tree and can be like a shrub in the early stages. **FLOWER** Yellowish-white flowers in terminal racemes. Ornamental varieties have striking purple

flowers. **LEAF** Distinctive butterfly-shaped leaves with two rounded lobes. Lighter green on undersides. Leaves arranged alternately on stalks. **FRUIT** Flat pod wrinkled. **HABITAT & DISTRIBUTION** Common in the dry lowlands. Can tolerate very dry environments. **ETYMOLOGY** Genus name honours the 16th-century herbalists John and Caspar Bauhin – pair of leaflets on leaf suggests brothers. *Racemosa* is a reference to the flowers being in racemes.

Indian Laburnum ■ *Cassia fistula*
(S: Ehela, T: Tiru kontai)

STRUCTURE Medium-sized tree with light brown trunk. Begins to branch very close to the ground and can look like a tall shrub. In botanical terms it has an untidy habit. When in flower, one of the most attractive flowering trees in Asia. Bark greenish-grey and smooth when young, and brown and rough when old. The laburnum *L. anagyroides* of Europe has pea-like flowers. **FLOWER** Very showy yellow flowers in long, pendulous racemes. Flowering often peaks to coincide with leaf fall, which makes the flowers even more prominent. Five petals with three stamens with long, upwards curving filaments and anthers opening along longitudinal slits. Four stamens have short filaments, and a further three are without pollen. **LEAF** Opposite pair of leaflets (pinnate) along midrib of leaf. Leaves can be about 30cm long with about eight pairs of leaflets. Elliptic leaflets have pointed tips. Leaves have short stalks. **FRUIT** Long, cylindrical, brown or black pod up

to 1m long, containing 40–100 seeds within pulp. Seeds separated by transverse partitions. Sloth Bears, jackals and monkeys eat the pods and play a role in dispersal. **HABITAT & DISTRIBUTION** Common roadside tree, and popular in parks and public spaces. **ETYMOLOGY** One explanation for the name relates to the bark of the tree being exported to Europe. It was rolled up into tubes

and this may have given rise to *fistula*, a Latin reference to pipe. Alternatively, *fistula* may refer to the pods. *Cassia* comes from the Greek name *Kassia*.

Horse Cassia ■ *Cassia grandis*

STRUCTURE Medium-sized tree with spreading branches and feathery look. Bark greyish. Rust-coloured, downy appearance to young leaves, stalks and twigs. Tree can be crooked, and in open spaces it branches early, spreading out to form a rounded canopy. **FLOWER** Numerous pink flowers in long racemes. Flowers February–April. **LEAF** Leaflets arranged oppositely along midrib with no terminal leaflet (paripinnate). Leaflets have parallel sides, are rounded at bases and tips, and are about 30cm long with around 10 pairs of leaflets. Young leaflets reddish. **FRUIT** Huge black cylindrical pod can be slightly compressed, and may reach 1m in length. Pulp between seeds has unpleasant smell. **HABITAT & DISTRIBUTION** Native to South America and the Caribbean, and widely introduced in the tropics as an ornamental tree. The *Cassia* genus has its greatest diversity in tropical America. **ETYMOLOGY** *Grandis* means large in Latin. *Cassia* comes from the Greek name *Kassia*.

Golden Cassia ■ *Cassia auriculata*
(S: Ranawara)

STRUCTURE Bush that can grow to about 2m tall, but typically reaches around 1m. One of the most visible and hence familiar plants in national parks, such as Yala in the dry lowlands. A community of these plants often grows together, and identification is helped by one or more of the bushes having conspicuous yellow flowers at any time of the year. Golden blooms add splash of colour to what can sometimes be a harsh, dry setting rendered in dust-sprinkled shades of brown. **FLOWER** Showy, stalked yellow flowers in panicles arising from axils. Sepals and petals yellow. Ten stamens, three of which are long, four medium and three short and infertile (staminodes). Dried flowers are used to make a herbal tea. **LEAF** Leaflets arranged oppositely (pinnate) with up to 13 pairs of leaflets. No terminal leaflet. Leaflets rounded ellipses. Linear gland on leaf midrib (rachis) between pairs of leaflets. **FRUIT** Seed pod flat and long. Seed chambers visibly articulated in pods. Ripened pods turn chocolate-brown. They carry a persistent style. **HABITAT & DISTRIBUTION** Widespread in the dry lowlands. **ETYMOLOGY** *Cassia* comes from the Greek name *Kassia*. *Auriculata* refers to an ear-like appendage.

Nam Nam ■ *Cynometra cauliflora*
(S: Nam Nam)

STRUCTURE Evergreen tree that under the right conditions can grow into a tall tree.
FLOWER Flowers in racemes arising from leaf axils, or on woody knobs on trunk. Typically white but sometimes with pinkish tinge.
Petals turned inwards at tips. Large number of protruding stamens, unusual for members of this family, a conspicuous feature. **LEAF** Leaflets about 15cm long and arranged opposite each other on brown midrib. No terminal leaf (paripinnate). Leaves elliptic, tapering to blunt tip. Young, pink leaves sheathed in papery scales. **FRUIT** Woody, short and wide pod. **HABITAT & DISTRIBUTION** Native to Assam and Malaya. Introduced to other tropical countries as an ornamental tree for its evergreen foliage and attractive flowers.
ETYMOLOGY *Cynometra* is derived from the Greek *kuon* (dog) and *metra* (womb).

Flamboyant ■ *Delonix regia*
(S: Mai Mal; T: Mayaram)

STRUCTURE Large tree that can be sprawling and many branched if allowed room to grow. Pale trunk and branches. Common roadside tree in towns and villages, instantly

recognizable when in flower. **FLOWER** Large red or orange flowers occur in clusters. Green sepals do not overlap each other. Ten stamens extend beyond petals. Upper petal has white or yellow stripe. **LEAF** Leaf midrib has opposite lateral leaflets (pinnate), which in turn have opposite pairs of pinnae (bipinnate or twice pinnate). Pinnae small and leaves overall give tree a feathery appearance. Leaves can be about 60cm long. **FRUIT** Pod about 30cm long, flat, with numerous seeds. **HABITAT & DISTRIBUTION** Believed to have originated in Madagascar, although no longer found anywhere in a wild state. Introduced plant popular in public spaces and on roadsides. Widely planted in moist tropical belt worldwide as an ornamental. **ETYMOLOGY** *Delonix* means evidently or obviously, from the Greek *delos* (evident) and *onux* (claw). *Regia* is Latin for royal.

Sickle Bush ■ *Dichrostachys cinerea*
(S: Andara)

STRUCTURE Untidy small, thorny bush. Typically waist high, but where conditions suit it, grows taller than head height, up to 6m. **FLOWER** Two-coloured, drooping flowering spike. Basal part attached to flower stalk pink with filaments, and apical part has short yellow stamens. Flowers are attractive but this being a common wayside plant, they do not draw much attention. **LEAF** Doubly pinnate, compound leaves that look feathery. **FRUIT** Typical legume pods that are contorted and twisted, and hang in bunches. **HABITAT & DISTRIBUTION** Found in the dry lowlands. One of the most common plants that visitors can see on safari in game parks in the dry lowlands.

Indian Coral Tree ■ *Erythrina variegata*
(S: Erabadu; T: Mullu murukku)

STRUCTURE Soft-wooded, medium-sized tree. Bark smooth and greenish-grey. In young trees bark has conical black spines. **FLOWER** Deep scarlet flowers, structured like pea flowers, densely packed into racemes. Unbranched lower part of inflorescence stalk (peduncle) can be more than 35cm long. Calyx has five tiny teeth. Upper petal (standard) is the largest, and wings and keel petals are distinctly smaller. Stamens project beyond petals (exserted). **LEAF** Each leaf comprises leaflets (trifoliate), rhomboid in shape. Two

are opposite each other with the third being a terminal leaflet. Leaf margins smooth (entire). Widest part of leaflet may exceed its length. Flowering can take place after leaf fall. **FRUIT** Large pod has about eight seeds and is jointed because of pronounced swelling around seeds. **HABITAT & DISTRIBUTION** Found in the lowlands. Widely planted in gardens and public areas. **ETYMOLOGY** Genus name is derived from the Greek *erythros* for the colour red, a reference to the flower colour.

Quickstick ■ *Gliricidia sepium*
(S: Weta Hiriya; T: Kona)

STRUCTURE Small tree with slender trunk and long, slender branches. Bark greyish, smooth or slightly fissured.
FLOWER White or pink, showy flowers, structured like pea flowers. Ten stamens, with nine joined to form sheath, and one separate (diadelphous). **LEAF** Midrib of leaf has opposite pairs of leaflets (pinnate). There is a terminal leaflet (imparipinnate).
FRUIT Blackish, wingless pod flat and without wings, opening into two leathery valves (dehiscent). Leaflets towards tip progressively larger. **HABITAT & DISTRIBUTION** Native to tropical America, and widely planted as a live fence in village gardens and commercial estates. Fast growing and also enriches the soil because it contains nitrogen-fixing bacteria.
ETYMOLOGY Genus name is derived from the Latin *glis* (dormouse or rodent) and *caedere* (kill) – a reference to the seeds being used as a rodenticide.

Gal Karanda ■ *Humboldtia laurifolia*
(S: Gal karanda)

STRUCTURE Small tree that can be mistaken for a bush at the early stages of colonizing a rainforest edge. Bark smooth and greyish-green. Drooping branches. **FLOWER** Inflorescence striking. Flowering raceme erect in contrast to drooping foliage. Honey-scented white flowers tinged with pink, and draw in bees. Flowers have five projecting stamens and petals; long stamens very conspicuous. **LEAF** Glossy green foliage. Leaves have five pairs of opposite leaflets (pinnate). No terminal leaflet (paripinnate). Tree provides an excellent example of the many ant and plant associations in the rainforest. Twigs have a hollow providing a home (domatia) for ants. Openings are located below leaf

stalks, and on careful inspection ants can be seen going in and out of them. **FRUIT** Pod flat and about 10cm long. **HABITAT & DISTRIBUTION** Native to Sri Lanka and south-western India. Very common in wet-zone rainforest patches. Six species in India and Sri Lanka, but only one recorded in Sri Lanka. **ETYMOLOGY** Genus name honours Alexander von Humboldt (1769–1859), the German naturalist and explorer, considered one of the greatest of his kind.

Mimosa-leaved Jacaranda ■ *Jacaranda mimosifolia*

STRUCTURE Medium-sized tree with feathery appearance, branching untidily. Bark
pale greyish-white. **FLOWER** Small calyx
has five teeth. Five petals form slender
tube with two lips. One has two lobes,
the other three. Four fertile stamens, two
long and two short. Single sterile stamen
(a staminode) longer than other stamens
and projects beyond tube. Purple flowers
grow in upright terminal panicles. Style
about the same length as stamens. **LEAF**
Opposite pairs of leaflets arranged along a
midrib, with these in turn being in opposite
pairs (twice pinnate or bipinnate). Latin
name alludes to mimosa-like appearance of
leaves. Leaves can be 35cm long. **FRUIT**
Rounded, flattened, woody capsule that
splits open to release seeds. **HABITAT &
DISTRIBUTION** Native to South and
Central America, and widely introduced in
the tropics as an ornamental. **ETYMOLOGY**
Jacaranda is a local Brazilian name.
Mimosifolia refers to mimosa-like leaves.

Jerusalem Thorn ■ *Parkinsonia aculeata*

STRUCTURE Grows as bush or small tree with brown or green bark. On first appearances at a distance, leaves appear grass-like. **FLOWER** Yellow flowers in clusters at bases of

leaves. Calyx cleft into five long, narrow segments. Petals broad with uppermost one having long claw. Ten stamens shorter in length than petals. **LEAF** Unusual leaf structure, on close inspection revealing very short midrib that divides into two branches at an acute angle to each other. On each are oppositely positioned, tiny leaflets. Leaf technically bipinnate, but appears very different from that of other bipinnate plants like the Flamboyant (see p. 72). Leaf stipules spiny. **FRUIT** Pod constricts between seeds, creating an undulating profile. It splits open on ripening. **HABITAT & DISTRIBUTION** Of three species in this genus, two are native to America and one to South Africa. This species from America is planted as a thorny hedge and can be seen in dry-zone villages. Look for it near the turn-off to Bundala National Park. **ETYMOLOGY** Genus name honours John Parkinson (1568–1620), an apothecary and author in London. *Aculeata* is Latin for thorny.

Yellow Flame Tree ▪ *Peltophorum pterocarpum*

STRUCTURE Tall tree with upwards sloping main branches. Smaller branches droop. Foliage has feathery look overall. **FLOWER** Numerous yellow flowers in panicles. Five roundish petals. Ten stamens, each separate. Calyx has five overlapping segments. Flowers quite showy, but often too high up above the traffic flow in cities like Colombo for people to see them. Around May–June roads may be carpeted with yellow where these trees are planted. **LEAF** Lateral leaflets arranged on opposite pairs to midrib and these in turn have opposite pairs of pinnae to form bipinnate leaves. No terminal leaflet. **FRUIT** Flat, reddish-brown pods show outlines of 1–4 seeds they contain. Pods winged on each edge. **HABITAT & DISTRIBUTION** Widely planted in towns and roadsides as a shade tree and as an ornamental for its flowers. **ETYMOLOGY** *Peltophorum* comes from the Greek *peltophoros* (shield bearer).

Indian Beech ■ *Pongamia pinnata*
(S: Magul Karanda; T: Punku – Sri Lanka, Ponga – India)

STRUCTURE Medium-sized tree forming dense canopy. Soft, smooth, greyish-brown bark. **FLOWER** Showy purple or white flowers. Ten stamens, with nine joined to form sheath and one separate (diadelphous). Flowers in clusters on long stalks from leaf axils. Each flower like a pea flower. **LEAF** Bright green leaflets arranged in opposite rows along midrib (pinnate), and terminal leaflet (imparipinnate) present. **FRUIT** Woody, thick pod with no wings. Single seeded, and pods take nearly a year to ripen. They do not split open, and for germination to occur the walls have to decompose. Watercourses that trees grow beside may help in dispersal of pods. **HABITAT & DISTRIBUTION** Native to India and Sri Lanka. Water-loving tree found beside waterways, including streams and ditches. Also widely planted as a shade tree. **ETYMOLOGY** *Pongamia* is derived from a Tamil name, 'Ponga', for the tree. *Pinnata* is Latin for feathered, a reference to the feathery leaves.

Andaman Redwood ■ *Pterocarpus indicus*
(S: Wal ehela; T: Kengei)

STRUCTURE Very tall tree with large canopy – some old street trees are like giants. When I grew up on Horton Place there was a very old, weathered, large tree across the road that to me, as a child, seemed to block out part of the sky. It had cavities in which Collared Scops-owls nested, and I imagined that it was the abode of goblins and evil spirits. Sadly, the ageing tree that had branches breaking off it became a hazard for commuters and it was taken down by the council. **FLOWER** Lightly scented, orange-yellow flowers form in terminal panicles, with several flowers on each panicle. Calyx has rounded teeth. Flowers pea shaped. Colombo's streets are sometimes carpeted in yellow flowers shed from this tree. **LEAF** Elliptic, shiny green leaves arranged alternately, and taper to a blunt point. Cluster of leaves on a twig will have a terminal leaf. **FRUIT** Tree can carpet the floor with round, winged seed pods. The wings aid dispersal by the wind. **HABITAT & DISTRIBUTION** Native to Malaya, and introduced in South Asia as a street tree. **ETYMOLOGY** *Pterocarpus* is derived from the Greek words *pteron* (wing) and *karpos* (fruit), in reference to the winged seeds. *Indicus* refers to being from India.

Rain Tree ■ *Samanea saman*
(S: Para Mara; T: Enak vakai)

STRUCTURE Large tree with some seasonal leaf fall, but in Sri Lanka it is not so pronounced as to see trees devoid of leaves. Many majestic examples of the tree can be seen on roadsides. In Colombo, Buller's Road has some especially fine examples lining the roadside and creating a shaded canopy over the road. Bark dark brown. During wet weather and at night, leaflets fold down, giving wilted appearance. **FLOWER** Flowers in solitary circular heads. Flower stalk (peduncle) long; around 12cm. Petals yellowish. Packed into flower centre are 20 red stamens that extend well beyond petals. **LEAF** Opposite rows of pinnae on a midrib (pinnate), which are in turn in opposite pairs (bipinnate or twice pinnate). Leaves about 40cm long. Leaflets towards tips of leaves are the longest, with the most pinnae. Midribs of leaflets and central midrib of leaf as a whole swollen at base. Pinnae blunt and dark green. **FRUIT** Pod lacks stalk (sessile). Pod straight and smooth

and about 20cm long. Seeds contained in pulp separated by transverse sections. **HABITAT & DISTRIBUTION** Native to Brazil, and widely introduced into tropical Asia in humid areas, where it is a fast-growing tree. **ETYMOLOGY** *Saman* refers to a local American name. English name Rain Tree comes from discharge of water droplets from a tree when it is infested by cicadas that discharge small droplets.

Asoka or Indian Saraca ■ *Saraca asoca*
(S: Asoka)

STRUCTURE Small tree with dense, rounded crown, and smooth brown bark. Young red leaves hang limply, interspersed with older dark green leaves. Mix of red and green a key characteristic. **FLOWER** Flowers very striking, despite lack of petals. Colourful calyx surrounded by coloured, spoon-shaped bracts. Flowers start off orange and yellow, and deepen to vermillion. Stamens slender and project beyond calyx. Flowers packed into corymbs. During the night flowers give out a scent. **LEAF** Opposite pairs of long leaflets (pinnate) on midrib, with no terminal leaflet (paripinnate). Smooth leaflets can be about 25cm long. **FRUIT** Flat, leathery pods that split open (dehisce) to release seeds. **HABITAT & DISTRIBUTION** Widespread from the lowlands to the mid-hills. Also planted in gardens and public spaces as an ornamental tree. **ETYMOLOGY** *Saraca* is a local West Indian name. *Asoka* means sorrowless.

Large-flowered Sesbania ■ *Sesbania grandiflora*
(S: Kathuru murunga)

STRUCTURE Small tree, feathery in appearance and attractive as an ornamental. Brown bark fairly smooth. Branches spread out, giving tree a rounded profile. **FLOWER** Flowers can be white, red or pink, and are on racemes arising from leaf axils. They are large and resemble pea flowers, but are more elongated. **LEAF** Small leaflets arranged in opposite pairs along central midrib with no terminal leaflet (paripinnate). Leaflet sides nearly parallel, with rounded tips and bases. There may be 20–60 leaflets on a leaf, which is about 30cm long. Each leaflet about 2–3cm long. **FRUIT** Long pod can be 30cm long. Slender pods slightly curved and four sided, with numerous seeds. **HABITAT & DISTRIBUTION** Native to Malaya and introduced to other parts of the tropics as an ornamental for its elegant appearance and flowers. **ETYMOLOGY** Genus name is derived from an Arabic name. *Grandiflora* is Latin for large flowers.

Tamarind ■ *Tamarindus indica*
(S: Siyambala; T: Puli)

STRUCTURE Large tree with fissured, dark brown bark. In young trees bark appears to have numerous vertical grooves. Large, horizontal, spreading branches. If not crowded in by other trees, branches close to the ground. Despite the feathery appearance, the canopy forms a dense, impenetrable layer, creating a deeply shaded area under the tree. This, together with the acidity of its leaves and fruits, results in the ground beneath often being devoid of other plants. **FLOWER** Flowers in loose clusters among leaves. They are unusual, with three unequal petals with two reduced in size. Flowers variably coloured in yellows and reds, with three staments, and on short stalks. **LEAF** Opposite pairs of leaflets arranged on central midrib. No terminal leaflet (paripinnate). Up to 20 pairs of leaflets, which are blunt on terminal ends. **FRUIT** Brown, cylindrical, swollen pod. Shiny seeds set within fibrous pulp. **HABITAT & DISTRIBUTION** Widespread in the lowlands. Especially common on roadsides and in village gardens as a shade tree, as well as for its fruits, which have many uses in cooking. **ETYMOLOGY** Tamarind is derived from an Arabic word referring to Indian Date. *Indica* refers to Indian.

RHAMNACEAE (BUCKTHORNS)

This family includes herbs and climbers, but mainly consists of trees and shrubs. There are around 55 genera with close to 1,050 species. It is a widespread family but absent from high latitudes, and most species are deciduous though some are evergreen. There may be spines on the plants' stems or on the tips of short branches. The simple leaves are usually arranged alternately but sometimes are opposite. The flowers are typically in axillary cymes, but the arrangement is highly variable – sometimes there may be a single flower, or the flowers can be clustered (fascicles) or in compound panicles. The flowers are radially symmetric (actinomorphic), and can be unisexual or bisexual. They have four or five free petals, and four or five stamens. The ovary may be superior to inferior, and the fruits are usually drupes, occasionally in the form of winged capsules.

Indian Jujube ▪ *Ziziphus mauritania*
(S: Masan, Debara)

STRUCTURE Can grow into a medium-sized tree, but many of the plants in the dry lowlands tend to take on the appearance of large bushes – most commonly in sandy soils, where water may be a limiting factor to growth. Bark very rough and can look gnarled and splintered. Thorny spines in pairs or singly on stems, below leaves. Branches spread out, often with a droop. **FLOWER** Small, hairy, greenish-yellow flowers are in axillary cymes without stalks (sessile) or in clusters on stems (fascicles). Petals turned down at edges. Stamens as long as petals (1–1.5mm). **LEAF** Leaves elliptic and three nerves strongly marked. **FRUIT** Fruit a round drupe, a fleshy covering with a hard stone inside. Green fruits turn orange or red when ripening. **HABITAT & DISTRIBUTION** Widely distributed from Australia, across South Asia, to West Asia. Similar species occur in Africa. In Sri Lanka abundant all over the dry lowlands. Also seen in the Colombo district (at Talangama Wetland, for example). Occurrence in wet zone may be due to accidental or deliberate introductions. **ETYMOLOGY** *Ziziphus* is derived from a Persian word, *zizafun*. *Mauritania* is a reference to north-west Africa. *Ziziphus mauritania* is considered a synonym of *Z. jujuba*. Jujube is derived from the French word *jujube*.

ULMACEAE (ELMS)

This family of trees and shrubs comprises seven genera with about 45 species. The family is largely one of the northern hemisphere, with species also found in the tropics across America, Africa and Asia. The plants' leaves are arranged alternately or oppositely, in a plane along the stem (distichous). A characteristic feature of the family is the asymmetrical leaf base. The leaves have teeth that are very pronounced in some species. The sap is watery. The flowers may be unisexual or bisexual, and are borne in axillary cymes or clusters on leaf axils. They are typically green or brown, with tepals that may be free or united. The number of stamens is equal to the number of tepals. The ovary is superior. The fruits can be drupes, flattened nuts or samaras.

Eastern Trema or Charcoal Tree ▪ *Trema orientalis*
(S: Gedumba; T: Mini)

STRUCTURE Usually grows as a small tree, but can also take the form of a shrub. Bark greyish-brown. Branches thin and straight. Trunk marked with stipular scars. **FLOWER** Tiny white flowers borne at bases of leaf stalks. Separate male and female flower clusters (cymes) occur on the same tree. Male cymes are compact.

LEAF Longish, pointed, finely toothed leaves arranged alternately in two rows. Underneath is silvery-white and upperside is rough. **FRUIT** Tiny black fruit. **HABITAT & DISTRIBUTION** Quick-growing, short-lived colonizer tree common in disturbed areas in the wet lowlands. **ETYMOLOGY** *Trema* is Greek for a hole, a reference to the pitted fruit. *Orientalis* means eastern in Latin.

MORACEAE (MULBERRIES, JACKFRUITS & FIGS)
This diverse family contains close to 1,200 species in nearly 40 genera. They include the well-known figs and conspicuous trees such as the Jackfruit and Breadfruit. Growth forms are varied and comprise perennial herbs, climbers, shrubs, and terrestrial and epiphytic trees. They typically have milky sap, and the leaf arrangement is alternate, but can be whorled or opposite. The leaves are often simple, but variations include deeply lobed leaves. The stipules usually wrap around the bud and leave a scar after falling. The flowering structures are in axillary racemes or cymes, which often form cup- or urn-shaped structures with the unisexual or bisexual flowers on the inside. In many species the fruits are fused together (as in the Jackfruit).

The incorrectly named 'rice paper' used in Japanese cuisine comes from the Kozo *Broussonetia kazinoki*, a member of this family. The Indian Rubber Fig was at one time an important commercial source of natural rubber. The Golden Fig is a common ornamental plant, which may be grown as a houseplant and grows as a large tree in gardens in Sri Lanka. The family also included the mulberries in the genus *Morus*, derived from the Greek word *moron* for mulberry. The family name *Moraceae* is also derived from the Greek word.

Jak or Jackfruit ■ *Artocarpus heterophyllus*
(S: Kos; T: Pla)

STRUCTURE Tall evergreen tree with straight trunk, dark brown bark and sticky sap. Generally compact, but occasionally with a few large, spreading branches. The timber is valuable for its lovely wood and termite-resistant properties. The trees are also important for the Critically Endangered Western Purple-faced Leaf Monkey. One of the largest fruiting structures of any plant. **FLOWER** Unusual flowers. Both male and female flowers occur on the same tree, arranged in a cylindrical head. Female flowers so small that they need to be viewed under a microscope to tell them apart. Male flowers bigger, and when mature their orange stamens can be seen. However, female head is bigger. **LEAF** Leaf a rounded oval. Young leaves dark green on uppersides and yellower on undersides. Leaves turn orange or red before being shed. **FRUIT** Tree exhibits cauliflory, with fruit growing directly on trunk or larger branches – an adaptation believed to help mammals access the fruits easily. For example, bats can land on the fruit directly, and are important in the dispersal of seeds. Composite or syncarpous fruit formed by many flowers pressed together on a common base. Seeds have a fleshy aril, 'waraka', which is eaten when ripe. It turns from colourless to orange. Unripe fruit cooked and eaten as a vegetable, and seeds also roasted and eaten. Green fruit turns yellow as it ripens. Leathery outer skin has prickles. **HABITAT & DISTRIBUTION** Introduced species, totally absent in areas of undisturbed forest, though one of the most widely planted trees – many Sri Lankans assume it is native. Grows well in warm, wet areas. Genus comprises typical rainforest trees. **ETYMOLOGY** *Artocarpus* is derived from the Greek words *artos* (bread) and *carpos* (fruit), and *heterophyllus* from the Greek words *heteros* (different) and *phullon*.

Breadfruit ■ *Artocarpus altilis*
(S: Rata-del; T: Era-pla)

STRUCTURE Medium-sized evergreen tree with spreading crown. Dispersion of branches is untidy. Bark can vary from greenish-grey to brown, with mottling or faint bands. **FLOWER** Male and female flowers on cylindrical heads. Male head 12–30cm long. Female head stiff and upright. **LEAF** Leaf deeply incised into several pairs of lobes with terminal lobe. Central midrib and lateral veins strongly marked. Leaves dark green when fresh. Ornamental aspect of leaves was a factor when tree was first introduced to other tropical countries. **FRUIT**

Large, round, complex fruits have leathery skin with hexagon shapes. These are derived from the flowers, which have a common base and are pressed together to form a composite (syncarpous) fruit. **HABITAT & DISTRIBUTION** Native to Malaysia, Indonesia and the Pacific Islands, and now widely planted in home gardens as an important source of food and a shade tree. **ETYMOLOGY** *Artocarpus* is derived from the Greek words *artos* (bread) and *carpos* (fruit).

The Curious Case of the Figs

Ficus trees bear the familiar edible 'figs', erroneously referred to as fruits. In botanical terms, figs are not fruits. What we see as the skin of a fruit is in fact the outer layer of a receptacle upon which flowers are borne. The flowers are inside the fleshy fig, and there are generally three types of flower – male, female and gall. The flowers are minute and not visible unless a fig is cut open. The male flowers have three or five sepals and 1–3 stamens, and the female flowers may have sepals or be cup-like, and have a long style. The ovaries develop into a minute fruit with a single seed inside. The gall flowers are like swollen female flowers and have a short, funnel-shaped style. They are sterile and the style provides a nest chamber for a fig wasp to lay its eggs. The gall flowers also contain 3–5 sepals or have a cup-shaped calyx.

Each fig tree species has a specific fig wasp (family Agaonidae) associated with it, which is necessary for pollinating the flowers. Some fig wasps can lay their eggs in a few closely related species, but most need just one specific species of tree. Figs can still ripen without the female flowers being pollinated and setting seed, but they cannot propagate naturally by seed. In some figs, all three flower types are not on the same fig. Those containing male and gall flowers will be on one tree, and those containing female flowers will be on another – the two fig types never occur on the same tree. The former can be thought of as 'gall trees' and the latter as 'seed trees'. The figs need to be cut open to identify the tree type.

Fig wasps are tiny animals only 1–2mm long and lack a sting. An adult female wasp enters a fig and lays its eggs on gall flowers. The male wasps hatch first and inseminate the still-developing females. The males then die, their role fulfilled (the males are blind or nearly blind, wingless and may only live for a few hours). The females mature and leave the fruit, taking with them male pollen that they inseminate female flowers with while they search for gall flowers. Their ovipositor reaches down the short style of the gall flowers and inserts an egg into an ovary. The female also injects a fluid into the ovary that stimulates it to grow, and the food that reaches the ovary for a plant embryo is actually absorbed by the fig-wasp egg. After laying their eggs in gall flowers, the female wasps die of starvation and exhaustion without having eaten anything since they hatched. The fig tree is now a surrogate mother for a clutch of fig-wasp eggs. If a fig species has separate gall trees, their figs are rendered distasteful by the fig wasps to reduce the risk of them being eaten before the new-born wasps leave the figs. The fig wasps hatch into tiny maggots that grow by eating the ovule, and the maggots hatch into male or female wasps to repeat the cycle.

Fertilized and ripened figs that have female flowers will have seeds that have set. Many seeds survive the passage through the guts of birds and mammals that eat them, and are dispersed away from the mother tree when they are passed out by animals.

The fig trees and fig wasps are interdependent on each other, and one cannot survive and propagate without the other. While a fig tree can be taken from its original location and planted in another, if the fig wasp is absent in the new location (for example in a different country), that tree will never bear seed. The fig tree and fig wasp interdependency is one of the most fascinating stories in botany.

Banyan ■ *Ficus benghalensis*
(S: Nuga; T: Al)

STRUCTURE Familiar roadside tree in India and Sri Lanka. Bark greyish and thin, peeling off in strips. Tree can grow to enormous proportions, typically beginning life as an epiphyte. Birds and bats drop seeds on plants and buildings, where the seeds germinate, sending down aerial roots to anchor themselves to the ground and obtain nutrition, then progressively throttling the host plant, which may even be a large tree. Large fig trees are almost like a mini ecosystem by themselves, and their fruits provide food for many animals. Branches spread widely and are almost horizontal. After anchoring to the ground, the aerial roots thicken into trunks. Sri Lanka does not match India for historic Banyan trees with vast canopies, which appear like a community of trees but are actually just one tree spreading out. **FLOWER** Not visible (see text box, p. 91). **LEAF** Leaves arranged alternately, and stiff (coriaceous). Mature leaves smooth; young leaves downy. Leaves large, oval to rounded, about 12cm wide, with stout leaf stalk (petiole), and pronounced, bulging veins. Leaf narrows slightly towards blunt tip. **FRUIT** Figs that are receptacles for flowers form in pairs on leaf axils. They lack a stalk (sessile). As they ripen, they turn from green to bright red, providing a clear visual signal to fruit-eating (frugivorous) animals. Male, female and gall flowers all in the same receptacle. **HABITAT & DISTRIBUTION** Native to India, and believed to be a naturalized species in Sri Lanka. Common throughout the lowlands. **ETYMOLOGY** *Benghalensis* means from Bengal, a region in India. English name Banyan believed to be derived from the name of a group of Indian traders, the Banyas, who apparently conducted their trade under the shade of the Banyan.

Golden Fig ■ *Ficus benjamina*
(S: Wal Nuga)

STRUCTURE Medium-sized tree. A strangling fig, though many trees have been cultivated and have grown from the ground upwards. Old trees have pillar roots. Bark smooth with horizontal lines. **FLOWER** Not visible (see text box, p. 91). **LEAF** Leaves have an elongated, oval shape, and pointed tips. Towards tip, leaf bends sideways a little, asymmetrically. **FRUIT** Yellow figs occur in pairs. Roundish in shape. **HABITAT & DISTRIBUTION** Found mainly in moist lowlands below 1,000m. Some fine old trees can be seen in cities such as Colombo. Also planted in the dry zone as an ornamental. Worthington records it as being introduced in 1861. **ETYMOLOGY** *Benjamina* is derived from the Sanskrit for Banyan. Note that the English name Banyan is also derived from the local word Banyan.

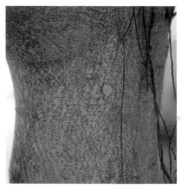

Bo or Peepal ■ *Ficus religiosa*
(S: Bo; T: Arachu)

STRUCTURE Large, spreading tree. At times rooted from the base, as when planted in temples. Under natural conditions can be epiphytic and drop aerial roots into the ground. Bark pale grey and flakes off. Widely branched, with no clear architecture to branch formation. **FLOWER** Not visible (see text box, p. 91). **LEAF** Pendulous leaves distinctive with long stalks, and triangular in shape with long drip tip. They catch the slightest breeze and shimmer in the wind. Uppersides darker than undersides, providing contrast of greens. **FRUIT** Figs form in pairs at bases of leaf axils, and ripen to purplish-black. They are popular with birds and bats, and are widely dispersed. New saplings often arise on tops of trees or buildings, and eventually strangle the host plant as they send roots down to the ground to draw nutrition to supplement what they take from the air and rain. **HABITAT & DISTRIBUTION** Native to India. Introduced and widely grown at Buddhist temples because of its religious association with the Buddha, who attained enlightenment while meditating under a Bo tree; every Buddhist temple has a tree as a focal point in its courtyard. **ETYMOLOGY** *Religiosa* in Latin means religious and refers to the tree's religious associations in Asia, especially with Buddhism.

India Rubber Tree ■ *Ficus elastica*

STRUCTURE Grows into a very large tree with smooth trunk and branches. Trunk buttressed, and bark can be reddish-brown or grey. Characteristic shiny leaves. A few aerial roots. **FLOWER** Not visible (see text box, p. 91). **LEAF** Elliptic leaves arranged alternately and are about 25cm long, stiff and reflective on uppersides, with pointed tips. Young leaves wrapped in pinkish sheaths, which are stipules at bases of leaf stalks. These long stipules fall after leaves have opened. Leaves have prominent veins that are parallel to each other and nearly perpendicular to midrib. Leaf similar to that of the Domba (see p. 48). **FRUIT** Figs form in pairs at axils of fallen leaves. The receptacles contain male, female and gall flowers. Figs egg shaped and turn greenish-yellow when ripe. **HABITAT**

& DISTRIBUTION Native to India. Introduced and planted as an ornamental. **ETYMOLOGY** *Elastica* refers to the plant yielding what was termed India rubber, which was made from the milky-white sap. The tree, native to Assam, Nepal and Myanmar, was at one time planted on a commercial scale for manufacturing rubber before the industry moved to *Hevea brasiliensis*, which was introduced from Brazil to other tropical countries.

Cluster Fig ■ *Ficus racemosa*
(S: Aththikka; T: Atti)

STRUCTURE Tall, many branched tree that grows rapidly. Smooth green bark. No aerial roots in this species, which is not a strangler fig. Grows with single main stem. Forms dense canopy with a few large branches and dense network of smaller branches. **FLOWER** Not visible (see text box, p. 91). **LEAF** Leaves small and arranged alternately. Distinct, thin leaf stalks. Young leaves toothed. **FRUIT** Green figs form in dense, branched clusters on trunk and branches (this is unusual as in most fig species, figs grow on leaf axils). Figs turn red on ripening, and are round with a short stalk. Ripe figs soft with skein of veins visible under skin. **HABITAT & DISTRIBUTION** Found mainly in the wet lowlands below 700m. **ETYMOLOGY** *Racemosa* refers to the raceme-like or branched cluster of figs. Raceme is a term usually applied to an arrangement of flowers. In figs, however, the flowers are not visible.

COMBRETACEAE (COMBRETALES)
Although this is mainly a tropical family of evergreen or deciduous lianas, shrubs and trees, it extends into the subtropics and warm temperate regions. There are about 500 species in around 20 genera. Many of the species are economically important for their timber. A characteristic of the family is the unique unicellular, compartmented hairs on the leaves, which are alternate or simple, with stipules absent or vestigial. The flowers occur in inflorescences and are typically radially symmetrical. They are bisexual in most species, but some are unisexual. Some species, such as the Indian Almond, have edible fruits.

Arjun or Kumbuk ■ *Terminalia arjuna*
(S: Kumbuk; T: Marutu)

STRUCTURE Large tree with pale, flaky bark. Associated with water and can form riverine forests. Trunk buttressed at base, and some old trees have enormous buttresses. Sheds leaves during dry season (deciduous). Flaking bark is a good identifying feature. **FLOWER** Small yellowish flowers on narrow spikes. Profusion of flowers decks tree when flowering is at its peak. Flowers emit an unpleasant smell. **LEAF** Long, bluntly pointed leaves arranged in pairs along branch, almost opposite each other. Leaf stalks very short, almost absent. **FRUIT** Green single-seeded fruit egg shaped or pear shaped overall, and narrowly winged along five sides. **HABITAT & DISTRIBUTION** Common waterside tree in the dry zone. **ETYMOLOGY** *Arjuna* is a local Indian name for the tree.

Bulu ■ *Terminalia bellirica*
(S: Bulu)

STRUCTURE Large tree with a few key branches that spread out to form dense canopy. Sheds leaves during dry season (deciduous). Trunk buttressed at base. **FLOWER** Flowers in simple, unbranched spikes, with male and hermaphrodite flowers mixed. Flowers small and greenish-yellow. They emit a strong, sweet, sickly smell. Woolly calyx has brown hairs on the inside. Petals almost absent. **LEAF** Large, elliptic leaves clustered at ends of branches. Leaf margins smooth (entire), and mature leaves smooth. Leaves about 25cm long, with leaf stalk (petiole) around 3cm long. **FRUIT** Small, velvety fruit can vary from round to oval in shape. Contains a single seed. Fruits popular with animals, and kernel can be eaten by humans. **HABITAT & DISTRIBUTION** Found in the dry lowlands and intermediate zone; also in rainforests. **ETYMOLOGY** *Bellirica* is derived from an Arabic name.

Indian Almond ■ *Terminalia catappa*
(S: Kottamba)

STRUCTURE Medium-sized tree that sheds its leaves in the dry season (deciduous). Bark rough and greyish-brown. **FLOWER** Interesting small whitish flowers in spikes arising from leaf axils, often high up. Upper flowers on spike are male and lower ones are hermaphrodite. Petals nearly absent. **LEAF** Large, elliptic leaves about 25cm long, with smooth edges (entire). Leaf narrow at base and widens out. Leaves arranged alternately, with short stalks (petioles). They turn orange and red before falling, reminiscent of autumnal colours of trees of high latitudes. **FRUIT** Green elliptic fruit slightly compressed with two ridges. Contains a single edible seed, encased in a tough, fibrous coat. **HABITAT & DISTRIBUTION** Native to Malaya, and widely planted in southern Asia. Seeds of plants near waterways may be transported by water and colonize new habitats. Largely absent away from human settlements. **ETYMOLOGY** Genus name *Terminalia* is derived from the Latin for terminal in reference to rosettes of leaves clustered at ends of branches. *Catappa* is a local name in Malaya.

Lythraceae (Pomegranates & Loosestrifes)
This family includes a wide range of growth forms, including herbs, shrubs and trees.
It has a worldwide distribution, with about 600 species in around 30 genera. The
plants' simple leaves are typically opposite, but can also be alternate or whorled. Many
species have glands on the bases of the leaf stalks. The secondary veins are arched to
form a vein running parallel to the edge of the leaf (submarginal vein). The flowers
can be solitary, axillary racemes or cymes, and in all but one genus they are bisexual.
Typically, there are 4–6 sepals, which form a tube that is persistent in the fruit. There
are generally 4–6 petals, but petals can also be absent. There can be 2–50 stamens,
usually of two or three different lengths. Plant watchers in Europe will be familiar with
the Purple Loosestrife *Lythrum salicaria*, which is a member of this family. *Lythrum* is
derived from the Greek word *lythron* (blood), a reference to the red flowers of plants in
the genus *Lythrum*.

Queen's Flower ■ *Lagerstroemia speciosa*
(S: Murutha; T: Pu murutha)

STRUCTURE Large tree, although trees in Colombo, for example, are never very big.
Guildford Crescent in Colombo is a good place to see it, as is the car park of the Beddegana
Wetland Park. A beautiful tree when in flower. Very important timber tree in north-
eastern India. **FLOWER** Purple or lilac flowers, borne on terminal panicles, have six petals.
Calyx divided into six triangular segments, and purple petals wrinkled. Stamens equal in
this species, but in others in this genus, six are longer and thicker than the others. Purple
stamens have yellow anthers. **LEAF** Leaves arranged in pairs opposite or nearly opposite
each other. Margins smooth (entire). Leaves have prominent veins. Short, stout leaf

stalk. **FRUIT** Round
brown fruit. Leathery
skin splits open from
3–6 valves to release
winged seeds. **HABITAT
& DISTRIBUTION**
Native tree found
in forest waterways.
Planted widely as
an ornamental.
ETYMOLOGY Genus
honours M. Lagerstroem
(1691–1759), a Swedish
patron of science.
Speciosa is Latin for
beautiful, a reference to
the flowers.

MYRTACEAE (MYRTLES, EUCALYPTUS & CLOVES)

This large family contains about 6,000 species in around 140 genera, many of which are familiar plants. The family is mainly tropical, but extends its range to the temperate zone. Most of the species grow as tall trees or shrubs. Their leaves are typically opposite, but also subopposite and spirally arranged, and generally simple without stipules. The leaves contain translucent gland dots with ethereal oils that when crushed are aromatic. The flowers occur in many forms, from inflorescences to solitary flowers, and have 4–5 sepals and petals, and many showy stamens. The fruits may be dry or fleshy.

Guava ▪ *Psidium guajava*
(S: Pera)

STRUCTURE Large shrub or small tree. Young branches downy. Grey bark with brownish flakes distinctive, giving rise to mottled and flaky appearance. **FLOWER** Flowers form on leaf axils and are on stalks (peduncles) about 3–4cm long. There may be 1–3 flowers in each axil, each with 4–6 broad white petals that are spread out. Dense mass of stamens in centre. **LEAF** Elliptic leaves arranged oppositely, end in a point and are crinkled. Prominent venation. Uppersides nearly smooth; undersides downy. Short leaf stalk (petiole). **FRUIT** Round or pear-shaped berry with many hard, tiny seeds inside it. Depending on the variety, flesh is white or pink with varying hardness. In young fruits flesh

is hard, softening as it ripens. Skin green, turning yellowish as it ripens. **HABITAT & DISTRIBUTION** Originated in Brazil and widely planted in the tropics for its fruits, which may be made into a jam. Popular tree in home gardens. **ETYMOLOGY** *Psidium* is believed to originate from the Greek word *sidion* (pomegranate). *Guajava* is the Spanish name, which may have West Indian origins.

Indian Blackberry ■ *Syzgium cumini*
(S: Ma-dun)

STRUCTURE Large tree with broad crown and drooping branches. Smooth bark pale, mottled with occasional darker patches. **FLOWER** White flowers densely packed together in cymes with long stalks. Flowers themselves lack stalk (sessile). Many long stamens packed together and project beyond petals. As a flower matures, petals that are joined fall off together, exposing clusters of long stamens. **LEAF** Small, elliptic leaves arranged oppositely on stalks of modest length. Leaves slightly stiff or leathery, reflective on uppersides, and with smooth margin and pointed tips. Leaves

prominently veined, and if held against the light translucent glandular dots can be seen. **FRUIT** Small, oval fruits, typically with single seed, turn from pink to black when ripe. Flesh juicy, and fruits are popular with birds. **HABITAT & DISTRIBUTION** Found in monsoon forests in the dry lowlands. Prefers sites near water. Some magnificent trees can be seen in national parks such as Wilpattu. **ETYMOLOGY** *Cumini* in Latin refers to cumin.

Jambu or Malay Apple ▪ *Syzygium malaccensis*
(S: Jambu)

STRUCTURE Small tree that may also grow as a small shrub. Smooth bark pinkish-grey with thin flakes. **FLOWER** Crimson flowers grow in clusters. Their densely packed, numerous long stamens are reminiscent of a brush. **LEAF** Oblong, deep green leaves arranged opposite each other. Leaf margins smooth (entire). Leaf stalks very short. **FRUIT** Soft, fleshy fruit (known as *jambu*) crimson in colour, edible and popular, and can be eaten without peeling. Pear shaped with remains of fleshy calyx segments at fatter end of fruit. Pulp inside white. Several seeds inside berry. **HABITAT & DISTRIBUTION** Native to Malaya and widely planted in home gardens for its fruits, as well as its attraction. **ETYMOLOGY** Genus name *Syzigium* is derived from the Greek word *suzugos* (paired).

MELASTOMATACEAE (MELASTOMES)

This pantropical family comprises about 5,000 species in around 170 genera, occurring as herbs, woody climbers, lianas, epiphytes, shrubs and trees. Two-thirds of the species are in the neotropics. In Sri Lanka, small shrubs from this family are common wayside plants. They can occur across a wide elevational gradient, from lowlands to cloud forests. A characteristic of the family is the venation of the leaves. The lateral leaf veins are parallel to the margins and converge at the apex. The flowers are showy and often in shades of purple. The leaves are opposite, and decussate (alternate opposite pairs are at right angles to each other). Several species show anisophylly (a pair of opposite leaves at a node with a pronounced difference in size or shape).

The family is divided into two subfamilies, the Kibessioideae and Melastomatoideae. The latter has seven tribes, of which the Melastomeae and Miconieae are pantropical. Some tribes are endemic to regions. *Clidemia hirta*, native to the Caribbean, and Central and South America, is an invasive species that has spread around the world.

Blue Mist ▪ *Memecylon umbellatum*
(S: Kora Kaha; T: Pandi kaya)

STRUCTURE Small, bush-like tree up to 3–4m tall. Bark dark, looking blackish. A gorgeous tree when in flower. **FLOWER** Bright blue flowers hang in clusters, on branches in umbels. Flowering occurs in March and September. Filaments of anthers and styles blue but sometimes pink. Petals usually blue but may be pink. Blue and pink flowers in some trees. **LEAF** Simple, elliptic to oval leaves. Leaves stiff and veins often not visible. **FRUIT** Small berry, about 2 cm long. Yellow-green and ripening to black. **HABITAT & DISTRIBUTION** Confined to southern India and Sri Lanka. Fairly widespread in Sri Lanka from the lowlands to the mid-hills. Not conspicuous in wet zone. In dry zone, in places like Yala National Park, avenues of the tree lining the safari roads become prominent when in flower. At other times these wild trees are easily overlooked. **ETYMOLOGY** *Memecylon* is derived from the Greek word for the fruit of the strawberry-tree *Arbutus*. *Umbellatum* references the flowers being in umbels.

ANACARDIACEAE (CASHEWS)

The cashew family is widely distributed especially in the tropics, with a few species extending into temperate latitudes. It comprises around 860 species in two subfamilies (Anacardioideae, 60 genera, and Spondiadoideae, 23 genera), with a wide range of growth forms, from vines and shrubs, to trees. Many species contain sap that when exposed to air turns black. The leaf arrangement can vary but most species have alternate leaves. Most have flowers in an axillary inflorescence. The flowers are generally radially symmetric (actinomorphic), though in just a few species they may have bilateral symmetry (zygomorphic). The plants can be unisexual or bisexual. The flowers typically have 3–5 sepals fused at the base, and 4–5 petals, but may have up to eight petals. There are twice as many stamens as petals, inserted on the margin of the disc. The ovary is superior.

Cashew Nut ■ *Anacardium occidentale*
(S: Caju; T: Mundiri-maram)

STRUCTURE Small tree with canopy that spreads widely. **FLOWER** Flowers are in panicles. White petals turn red. Petals are in the form of thin straps (ligulate). **LEAF** Broad, oval, corrugated leaves with prominent midrib and lateral veins. **FRUIT** Fruit comprises swollen stem that is edible, as well as claw-like fruit containing edible nut.

HABITAT & DISTRIBUTION Native to South America and the West Indies, and introduced widely into the tropics as a cash crop. Naturalized in Sri Lanka and found in the lowlands. Occurs in the suburbs of Colombo due to the presence there of cashew plantations until comparatively recent times. **ETYMOLOGY** *Anacardium* is derived from the Greek words *ana* (up) and *kardia* (heart). *Occidentale* is Latin for western.

Hik ■ *Lannea coromandelica*
(S: Hik; T: Odi)

STRUCTURE Medium-sized, untidy tree with straggling branches and smooth greenish bark. Leaf fall may be pronounced during the dry season – it is less pronounced or almost absent in trees planted in the wet zone. **FLOWER** Trees can have male and female flowers on separate trees (dioecious), or male and female flowers on the same tree (monoecious). Male flowers have 8–10 stamens. Female flowers have three or four styles. Greenish flowers small and on racemes at ends of branches. Female racemes simple; male racemes multi-branched (compound). **LEAF** Opposite 4–5 pairs of pointed leaflets along midrib with terminal leaflet (imparipinnate). Leaves about 45cm long. Leaflets have smooth margins (entire), and short stalks (petiolulate). They are clustered at ends of branches. **FRUIT** Oval-shaped berries small and flat, and found on female branches or female trees. **HABITAT & DISTRIBUTION** Widely planted as a hedge tree. The branches are full of starch, which makes the tree easy to propagate by planting cuttings. **ETYMOLOGY** Genus name is derived from a local name in Senegambia for one of the species in the genus.

Mango ■ *Mangifera indica*
(S: Amba; T: Manga)

STRUCTURE Medium-sized, much-branched tree. Branching haphazard. Evergreen, dark green leaves stand out in dry parts of Sri Lanka when deciduous trees have shed their crowns. Bark dark brown and rough. **FLOWER** Tiny yellowish flowers on hairy panicles. Flower stalks tiny to absent. Flowering panicles can take different forms. There can be separate male or female flowers (monoecious), a mix of male and female flowers, or bisexual flowers only, or bisexual flowers may be present with male or female flowers (polygamous). Calyx and petals number 4–5. Each flower has 4–5 stamens, as many as the petals. Only two stamens are fertile with one or two anthers on stamen. **LEAF** Stiff and reflective (coriaceous) leaves arranged alternately. They are undivided and margins vary from smooth to slightly crinkled. Leaves about 25cm long and crowded at ends of branches. Long, narrow leaves taper to a point and droop slightly from short stalks. They emit a resinous smell when rubbed. Young leaves pinkish-brown or reddish and hang limply. **FRUIT** Familiar mango fruit is in botanical terms a single-stoned, fleshy berry. Stone covered in edible fibrous pulp that is much desired by mammals, including humans. Depending on the variety, smooth skin will turn from green to orange or red when ripe. End opposite fruit stalk narrower, with one of the long sides being convex, creating the 'mango' shape. **HABITAT & DISTRIBUTION** Widely planted as a shade tree and for its fruit. Also commercially farmed for fruits. Most trees probably modified to some extent from an original native species. **ETYMOLOGY** Genus name *Mangifera* is derived from mango and *fera* in Latin to mean 'to bear'. Hence *Mangifera* means to bear mangoes. 'Mango' is derived from the Tamil name.

Hog-plum ■ *Spondias dulcis*
(S: Ambarella)

STRUCTURE Evergreen tree that can grow quite tall, to a medium-sized tree. Branches spread out fairly neatly around main trunk. Bark pale grey and smooth. **FLOWER** Tiny yellowish-white flowers grow in clusters at ends of branches. **LEAF** Leaflets arranged in opposite (pinnate) or nearly opposite pairs along midrib. Terminal leaflet present (imparipinnate). Pinnate or nearly pinnate leaves arranged alternately on stout stalks. Leaves about 30cm long with 13–15 leaflets. Leaf tip pointed and margins finely saw toothed. **FRUIT** Green fruit has smooth skin and is like a fat egg. Turns yellow on ripening. Inside a white fibrous pulp is a hard stone with 1–3 seeds. Flesh edible when ripe. Bitter rind often removed. Slightly raw fruits are eaten with salt and chilli. **HABITAT & DISTRIBUTION** Originated in the Society Islands in the South Seas. Widely cultivated in the tropics for its fruits and as a garden tree for its neat appearance. **ETYMOLOGY** *Spondias* is a name used by Theophrastus. *Dulcis* in Latin means sweet.

BURSERACEAE (FRANKINCENSE & MYRRH)

Most members of this family are shrubs, with a few growing as climbers or epiphytes. It contains about 20 genera and nearly 800 species, and extends around the world in the tropics. A key feature of the family is the white or colourless sap, which is heavily scented. The flowers are radially symmetrical, and male and female flowers are often on separate plants (unisexual), though some species are bisexual. Close up to a light source, translucent glandular dots (pellucid) can be seen in many species. The flowers are typically in racemes, but other forms of inflorescence also occur. There are 3–5 sepals, typically partially fused. They persist in the fruit and continue to grow (accrescent). The stamens are in one or two whorls, and there may be as many or twice the number of petals. The female flowers also have stamens, but these are sterile (staminodes). The single style has a stigma with as many lobes as carpels in the ovary. The family name honours Joachim Burser (1583–1639), a German physician and botanist.

Ceylon Almond ■ *Canarium zeylanicum*
(S: Kekuna; T: Pakillipal)

STRUCTURE Large tree with numerous branches that are long and thick at the bases. Prominent buttresses and roots. Bark mottled and scaly. An injured tree exudes a clear resin. **FLOWER** Green flowers on terminal panicles. Male flowers densely clustered together, and have six stamens and three petals. Female flowers are few, and have short

styles. **LEAF** Leaves narrowly elliptic with pointed tips. Lateral veins prominent and contrast with greener leaves. **FRUIT** Longish, oval fruits have edible seeds. An oil is extracted from them for lighting. **HABITAT & DISTRIBUTION** Endemic tree mainly confined to wet zone in south-west up to the foothills; also recorded in dry zone. **ETYMOLOGY** *Canarium* is derived from a local Moluccan name. *Zeylanicum* refers to it being described as from Ceylon.

SAPINDACEAE (LYCHEE, MAPLES & HORSE CHESTNUTS)

This family is found mainly in the tropics and southern latitudes, but also includes temperate species previously regarded as part of the Aceraceae and Hippocastanaceae. The family is best known for its trees, but also includes woody climbers and shrubs. In the northern temperate zone, species are mainly in the genera *Acer* and *Aesculus*. There are about 2,000 species in around 150 genera. India has around 37 species and the Malesian region about 235. The plants' leaves are alternate or opposite. They take a variety of forms, from pinnate to palmate, with different numbers of lobes.

Fern-leaf Tree ▪ *Filicium decipiens*
(S: Pihimbiya; T: Chittira vempu)

STRUCTURE Small to medium-sized evergreen tree with leaves reminiscent of ferns. Bark dark, and reddish-brown where it has flaked off. Widely planted along roadsides, and may be confused as one of the many introduced ornamental trees. **FLOWER** Tiny white flowers arise in branching clusters in leaf axils and are easily overlooked. Flowers in February. **LEAF** A distinctive feature of this plant is the winged midrib, on either side of which opposite pairs of leaflets are arranged. No terminal leaflet. Leaflets lance-like, tapering into points at both ends. **FRUIT** Egg-shaped
berry containing a single seed.
Berry turns purple when ripe.
HABITAT & DISTRIBUTION
Native to western India and Sri
Lanka, as well as to eastern Africa.
ETYMOLOGY *Filicium* is derived
from filicia for ferns, a reference to
the shape of the leaves. *Decipiens*
means deceptive in Latin,
referring to the difficulty botanists
had in classifying the tree.

Ceylon Oak ▪ *Schleichera oleosa*
(S: Kon; T: Puvu)

STRUCTURE Tall, large tree with dark greyish bark and dense canopy. Sheds leaves during dry season. Trunk has small buttresses (fluted) at base. **FLOWER** Minute yellowish-green flowers lack petals, and are arranged in clusters arising from the same point (fascicles) on a raceme. Racemes up to 15cm long. Six to eight stamens. Style cleft into three or four. Some trees have only male flowers, while others have bisexual (hermaphrodite) flowers. **LEAF** Opposite pairs of leaflets on a midrib (pinnate). No terminal leaflet (paripinnate).

Leaves can be about 40cm long. Emergent leaves reddish, turning to dark green on maturity. **FRUIT** Round fruit smooth and about 2.5cm long. **HABITAT & DISTRIBUTION** Monsoon forests in dry and intermediate zones; also widely planted in cities such as Colombo. **ETYMOLOGY** Genus name honours the Swiss botanist J. C. Schleicher. *Oleosa* is a Latin reference to being rich in oil.

■ CITRUSES ■

RUTACEAE (CITRUSES)
This familiar family of shrubs and trees includes the Lemon, Orange and Grapefruit, cultivated extensively for their fruits. There are about 900 species in 150 genera, found mainly in tropical and warm regions. Many plants have aromatic leaves that when crushed give rise to a 'citrus' fragrance. The family is especially diverse in Australasia.

Ankenda ■ *Acronychia pedunculata*
(S: Ankenda)

STRUCTURE Small to medium-sized tree with straight trunk. Branches profusely at regular intervals. A colonizer species that grows on cleared land. Bark brown and smooth. Small lenticels on trunk. **FLOWER** Greenish-white flowers in axillary inflorescences. **LEAF** Oval-shaped leaf tapering to point at both ends. Crushed leaf smells strongly of turpentine. Fresh leaves bright green, fading to yellowish with age. **FRUIT** Greenish, globular fruits on long stalks. **HABITAT & DISTRIBUTION** Found in wet zone from the lowlands to the hills. Absent at high elevations.

Bael ■ *Aegle marmelos*
(S: Beli; T: Vilam)

STRUCTURE Small to medium-sized tree. Thorns arise on leaf axils. **FLOWER** Relatively large, greenish-white, sweetly scented flowers in open clusters along branches. **LEAF** Each leaf comprises three – or occasionally five – oval, pointed leaflets. Leaf stalk long. Leaves arranged alternately. Leaf margin smooth or crenate. **FRUIT** Round, greyish-green fruits, turning yellowish as they ripen. Wild varieties have small fruits about 8cm in diameter. Rind a tough, woody shell. Pulp orange-yellow and sweet. **HABITAT & DISTRIBUTION** Native tree found naturally in the lowlands; also planted in home gardens for its fruits. Plant has various medicinal uses. **ETYMOLOGY** *Aegle* was one of the water nymphs in ancient Greek mythology. *Marmelos* is the Portuguese name for the tree.

Lime ■ *Citrus aurantifolia*
(S: Dehi)

STRUCTURE Typically grows as a much-branched, thorny shrub, but can take the form of a small tree. Evergreen with glossy green leaves. Branches and twigs have numerous spines. **FLOWER** The same plant can have male and female flowers (monoecious) or bisexual flowers. Unisexual flowers will not have fully developed stamens or style. Flowers in small clusters in leaf axils. Calyx can be shallowly toothed. There are 4–5 waxy, white petals with 20–40 stamens. Number of stamens higher than in other citrus species, and they surround a style of about the same length. **LEAF** Leaves egg shaped (ovate). Margins finely toothed; teeth can be sharp or blunt. Leaf stalk winged; a trait shared with many citrus species. Leaf tip blunt. **FRUIT** Fruit like a sphere that has been pulled out at the two ends. Some variability in shape. At end opposite leaf stalk, fruit has a little projection. Fruit green, turning yellow as it ripens.

Typically 10 segments with several seeds. Smooth skin. Rind thin but tightly attached to fruit and does not peel off as easily as that of an orange. Pulp emits well-known citrus scent. **HABITAT & DISTRIBUTION** Native to Malesian region. Widely planted in Sri Lanka. **ETYMOLOGY** *Citrus* is derived from the Greek word *citron*. *Aurantifolia* in Latin refers to having orange-like leaves.

Wood Apple ■ *Limonia acidissima*
(S: Divul; T: Vila)

STRUCTURE Medium-sized, thorny tree with corrugated brown bark. Branches spread out fairly horizontally from straight, small trunk. Elephants are fond of the fruit and play a role in dispersing the seeds – these survive passing through their intestinal tracts. **FLOWER** Flowers white or pale green, sometimes with red tinge. In racemes or panicles growing out of leaf axils. Both male flowers (staminate or stamen-bearing flowers), and bisexual or perfect flowers (having both male and female parts). **LEAF** Stiff leaflets arranged in opposite pairs (pinnate) with terminal leaflet. Leaflet tapers to acute angle at base and has rounded tip to form elliptic shape. Leaf stalks between leaflets winged. Seen up close, leaves are dotted with glands. Crushed leaf has a faint smell of lemon. **FRUIT** Spherical fruit about

the size of a tennis ball or bigger, hanging from stalk. Within a pale (almost white) hard shell is a soft pulp with seeds embedded in it. Ripe pulp and seeds edible. Pulp used to make a popular fruit drink as well as jam. **HABITAT & DISTRIBUTION** Native to India and Sri Lanka. Common tree in the dry

lowlands. Dominant tree in semi-open scrub forests of national parks such as Yala and Wilpattu. Also widely planted in home gardens and commercial plantations in dry zone for its fruits. **ETYMOLOGY** *Limonia* is the Latin name for the citron, possibly a reference to the lemony smell of a crushed leaf.

Curry Leaf ■ *Murraya koenigii*
(S: Karapincha; T: Karivempu)

STRUCTURE Small tree with dense, shady crown, perceived as a bush because of its low height, but not a bush as it has a single trunk. **FLOWER** Flowers grow in terminal panicles and form rounded head (corymbose) of flowers. White flowers somewhat bell shaped. Ten stamens. Almost the whole tree is downy. **LEAF** Leaflets arranged in opposite pairs with terminal leaflet (imparipinnate). Lance-like leaves can be long, about 30cm. Leaflet margins have rounded teeth (crenulate). Crushed leaves strongly scented and used heavily in cooking for their aroma and flavour. **FRUIT** Fruit nearly round to egg shaped. Greenish-white turning purplish-black on ripening. End away from stalk has a point. **HABITAT & DISTRIBUTION** Native plant found in monsoon forest edges. Widely planted in gardens for curry leaves. **ETYMOLOGY** Generic name honours J. A. Murray (1740–1791), a pupil of Carl Linnaeus. The specific epithet honours J. G. Koenig (1728–1785), also a pupil of Linnaeus.

MELIACEAE (MAHOGANIES)
Found throughout the tropics and subtropics, this family comprises about 550 species in around 50 genera. The plants grow mainly as shrubs and trees, rarely as herbaceous shrubs. The family contains some important timber trees in the genera *Swietenia* and *Khaya*, which are widely cultivated for commercial timber. The plants' leaves are alternate or spiral, usually pinnate and rarely bipinnate. The radially symmetrical flowers are functionally unisexual. The inflorescences are cymose panicles and show much variation. They can be terminal or appear on the trunk or branches, or in leaf axils. The sepals and petals may be united or distinct. The fruits may be capsules, drupes, berries or rarely nuts. The Margosa, known as the Neem in India, has been used in Ayurvedic medicine for more than 2,000 years.

Margosa or Neem ■ *Azadirachta indica*
(S: Kohomba; T: Vepa)

STRUCTURE Medium-sized, much-branched tree. Rough brown bark in older trees can form ridges of longitudinal scales. One of the most common planted trees in the dry lowlands, often placed beside roads. Canopy provides good shade and drooping leaflets are quick to catch the wind. In Yala, cubs and subadult Leopards can be seen resting in

Margosa trees. **FLOWER** Sweetly scented white flowers are on cymes on long raceme, forming a thyrse. **LEAF** Oval leaf has serrated edges and tapers to a point. Leaf curves at terminal end. Leaflets in opposite pairs (pinnate) on leaf stalks that can have eight or more pairs. **FRUIT** Oval fruit turns yellow with ripening. **HABITAT & DISTRIBUTION** Native to Sri Lanka, India and Malaysia. Widely planted in Sri Lanka as a shade tree. Highly valued on Indian subcontinent for its medicinal properties. **ETYMOLOGY** *Azadirachta* is derived from the Persian word *azad-darakht*. *Indica* refers to India.

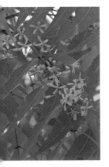

Indian Lilac or Ceylon Mahogany ■ *Melia azedarach*
(S: Lunumidella; T: Malai vempu)

STRUCTURE Medium-sized deciduous tree with greyish bark. Trunk fissured with long, shallow ridges. **FLOWER** Lilac flowers grow in branched racemes near bases of compound leaves. Stalks of flower racemes shorter than leaf stalks, and flowers can be hidden under foliage. Flowers emerge before new leaf growth, which follows soon after. Flower has 5–6 lilac petals surrounded by a tube formed by stamens. Emits a pleasing scent that is strongest at night. **LEAF** Basal area of leaf midrib has opposite pairs of leaflets. Each leaflet in turn has an opposite pair of pinnae (bipinnate) with a terminal pinna. At the tip of the leaf midrib is a single pinna. Near the tip, one or more opposite pairs of pinnae will arise from the midrib. Technically speaking these 'pinnae' at the apical end are 'leaflets', but simple leaflets, not compound leaflets as in the basal half of the leaf midrib. **FRUIT** Smooth, oval berries, shining green at first and ripening to yellow. **HABITAT & DISTRIBUTION** Originally a native of Baluchistan. Introduced to various tropical countries and has naturalized in warm climates. In Sri Lanka, also planted in wet zone. **ETYMOLOGY** *Melia* is the ancient Greek name for the Ash *Fraxinus excelsior*, with a leaf similar to that of the Margosa (see opposite), previously placed in this genus. *Azederach* refers to the local name for this tree in Iran.

Large-leaved Mahogany or Honduran Mahogany
■ *Swietenia mahagoni*

STRUCTURE Grows into a tall, straight-boled tree that is much prized for its timber. Tree is evergreen. Dark brown bark has strips flaking off it. **FLOWER** Greenish-yellow flowers are in panicles arising from leaf axils. The five petals spread outwards. Stamens form tube with 10 tiny teeth. Cherry-coloured disc at base of stamens. **LEAF** Opposite pairs of curved leaflets arranged on midrib without terminal leaflet (paripinnate). Leaflets narrowly elliptic and taper to a point. Leaflet stalk (petiolule) very short and not obvious. Leaves glossy green. Veins arising from central midrib towards margin very prominent. **FRUIT** Long, woody, conical brown fruits that are five celled, and about 8cm in diameter with many winged seeds inside. **HABITAT & DISTRIBUTION** Native to the West Indies and Central America. Widely planted for its much sought-after timber. In Sri Lanka found in commercial plantations, and occurs as scattered trees that have been grown to provide shade. **ETYMOLOGY** Genus name honours Gerard van Swieten (1700–1772), a Dutch botanist. *Mahogani* is a local West Indian name.

BIXACEAE (ANNATOO)
This neotropical family comprises one genus of five shrub or small tree species. The sap in their stems and leaves may be yellow, orange or reddish. *Bixa orellana* is widely cultivated in the tropics and may not even be found in the wild where it originated. The plants' leaves have long petioles, are entire and simple, and can fall off early or prematurely. The veins are palmate. The flowers are radially symmetric and showy in thyrsoid inflorescences. They have five sepals and five petals. The fruit is a capsule that splits longitudinally along the dorsal sutures of the wall (loculicidal). The plants' economic use chiefly involves extraction of the reddish dye annattoo, which contains mainly bixin.

Yellow Silk Cotton Tree ■ *Cochlospermum religiosum*
(S: Kinihiriya; T: Konga)

STRUCTURE Medium-sized to large deciduous tree. Branches short, stout and spreading. Bark pale and marked with longitudinal fissures. **FLOWER** A few bright yellow flowers borne on terminal panicles. Five sepals and five spreading petals. Many yellow stamens in centre of flower. Flowering occurs after leaf fall. **LEAF** A number of lobes radiating from centre of leaf where leaf stalk joins it. Leaf stalk not on the same plane as leaf lobe. This is different from digitate leaves, where leaflets radiate from basal point where leaf stalk and leaflets are all in one plane. Leaves at ends of branches and arranged alternately. Leaflets oval and pointed at tips. Leaf stalk fairly long at about 10cm. **FRUIT** Round or pear-shaped capsule with five cells and many seeds. Kidney-shaped seeds have cottonwool-like white threads attached to them. **HABITAT & DISTRIBUTION** Native to India. Introduced tree planted in temples and gardens. **ETYMOLOGY** Genus name is derived from the Greek *kochlos* (spiral shell) and *spermum* (seed).

DIPTEROCARPACEAE (DIPTEROCARPS OR MARANTIS)
Maranti is the name used for many of the dipterocarp species found in Peninsular Malaysia. In the 1970s they were a significant contributor to Malaysia's GDP, before the discovery of petrochemicals. This is predominantly an Asian family with its greatest diversity in Borneo. The trees originated on the South Asian plate and entered mainland Asia when the Indian tectonic plate, with India and Sri Lanka, made contact with Southeast Asia, before finally ending up where it is now.

This amazing family quite rightly has entire books dedicated to it. It is made up of two subfamilies. The Monotoideae comprise three genera and about 35 species. They are also found in northern South America (genus *Pseudomonotes*) and in tropical Africa (genera *Marquesia* and *Monotes*). The subfamily Dipterocarpoideae comprises 13 genera with about 840 species. The genera *Shorea*, *Hopea*, *Dipterocarpus* and *Vatica*, all of which are present in Sri Lanka, contain the most species. The taxonomy of the dipterocarps has been the subject of much debate, with some early authorities questioning whether the species in South America really belonged in this family. The debate will continue, but molecular genetics are bringing new insights.

Sri Lanka has 51 dipterocarp species in seven genera, namely *Cotyleobium*, *Dipterocarpus*, *Hopea*, *Shorea*, *Stemonoporus*, *Vateria* and *Vatica*. The country has lost many of its lowland rainforests due to a human population that has increased from under a million in the 19th century census to more than 22 million in modern times. Nevertheless, the fragments of rainforest that remain in locations such as Sinharaja, Morapitya and Kanneliya are wonderful places in which to see these forest giants.

Ceylon Dipterocarp or Hora ■ *Dipterocarpus zeylanicus*
(S: Hora)

STRUCTURE Very tall forest giant with straight trunk. This made it, like other dipterocarps, a favourite of foresters looking to harvest timber. Sparsely branched. Bark orange-brown, with large flakes peeling off, and little excrescences (like speckles) on surface (verrucose). **FLOWER** Flowers hang from racemes that are clustered in dense masses. Fifteen stamens. **LEAF** Oval-shaped, stiff leaf slightly wavy at edges. Lateral veins on leaves clearly marked. Leaves clustered around twigs. Leaf stipule has base growing around to opposite side of stem to embrace it (amplexicaul), leaving pale scars after leaf and stipule have fallen. Stipules on young leaves long; about 13cm and pinkish-red. **FRUIT** Young fruit purple. Two long wings around fruit aid dispersal. **HABITAT & DISTRIBUTION** Endemic to Sri Lanka and confined mainly to the wet lowlands of the south-west, below 1,000m. Also found around Moneragala in Uva Province, and on lower slopes of Central Province (e.g. Knuckles). Although isolated trees can still be found in random locations, stands are confined to protected areas. **ETYMOLOGY** *Dipterocarpus* is derived from the Greek words *di* (two) and *pteron* (wings), a reference to the two wings on the seed. *Zeylanicus* refers to it being described from Sri Lanka.

Hal ■ *Vateria copallifera*
(S: Hal)

STRUCTURE Tall forest giant growing with straight trunk. Trunk buttressed. Bark smooth and pale. Faint horizontal markings on trunk. Scaly pieces of bark are shed, leaving scalloped depressions. **FLOWER** Scented, cream-coloured flowers grow in panicles. Flower

stalks (pedicels) short. Forty-five to fifty-five stamens. Yellow anthers. Flowers around April–May. **LEAF** Leaf long and oval, and strong. Young leaves have pinkish-brown flush. **FRUIT** Brown fruit on long stalk. No wings. Seeds edible. **HABITAT & DISTRIBUTION** Endemic to Sri Lanka and widespread in lowland rainforests in south-west below 700m. **ETYMOLOGY** Genus name honours Abraham Vater (1684–1751), a German physician, anatomist and botanist.

> **MALVACEAE (MALLOWS)**
> As many as 1,500 species comprise this family, which includes the well-known garden flower, the 'Shoe Flower', or Hibiscus – also the name of one of the 75 genera in the family. The mallow family occurs in a variety of forms, including herbs, shrubs and trees. Mallows have five petals and clusters of stamens on a single stalk.

Baobab ■ *Adansonia digitata*
(S: Aliya Gaha; T: Papparapulli)

STRUCTURE Distinctive tree with disproportionately wide trunk. Bark grey and generally smooth, but wrinkled at bases of limbs. Colour reminiscent of an elephant's skin. Crown of tree looks proportionately small and is mushroom shaped. When the tree is bare of leaves, it looks as though it has been thrust into the ground with part of the trunk and roots sticking out, hence the name 'upside-down tree'. **FLOWER** Solitary flowers arise from leaf axils and hang from long stalk (pedicel). Calyx cup shaped and cleft into five. Five white petals. Hanging below them is a short, cylindrical staminal tube with numerous filaments tipped with anthers. **LEAF** Leaves have 5–7 long lobes radiating from end of a stalk. **FRUIT** Long, cylindrical fruit. Kidney-shaped brown seeds are within a pulp.

This is used to make a drink. **HABITAT & DISTRIBUTION** Native to Africa and believed to have been introduced by Arab traders. Largely confined to Mannar district, with large trees found on mainland Mannar as well as Mannar Island. A very large baobab about 1km away from the Mannar Fort is said to have been planted in 1477 by Arab sailors. **ETYMOLOGY** Genus name honours Michel Adanson (1727–1806), a French botanist. *Digitata* is Latin for 'fingered', in reference to the finger-like leaf lobes.

Trincomalee Wood ■ *Berrya cordifolia*
(S: Halmilla; T: Chavandalai)

STRUCTURE Very large tree with straight, tall trunk and large, strong branches. Bark smooth and brownish-grey. Wide, shade-giving canopy. **FLOWER** Numerous flowers in large terminal panicles. Calyx downy with irregular lobes. Five narrow white petals that spread out. Flowers have many free stamens that are about half the length of the petals. Yellow anthers. **LEAF** Heart-shaped (cordate) leaves arranged alternately and crowded at ends of twigs. Edges slightly wavy. Pointed leaf tip. Long leaf stalk. **FRUIT** Rounded capsule with persistent calyx. Capsule has six wings and three cells. Papery wings. Cells have 1–4 seeds. **HABITAT & DISTRIBUTION** Common native tree of the dry lowlands, prized for its timber. Large trees absent away from protected areas. **ETYMOLOGY** Genus name honours Dr Andrew Berry, a botanist who did much for the botanical garden in Calcutta, India.

Kapok or White Silk Cotton ■ *Ceiba pentandra*
(S: Kapok; Pulun Imbol)

STRUCTURE Tall tree with smooth, greyish-brown bark. Young trees characterized by spines on bark. Characteristic tree architecture with several whorls of more or less horizontal branches around trunk. Base of trunk typically buttressed. Deciduous tree with flowers emerging after leaf fall. **FLOWER** Bunches of creamy-white flowers (fascicles) arise on leaf axils. Flowers have bell-shaped calyx with five pointed teeth. Five petals downy on outside but shiny within. Five long stamens have 2–3 orange or brown anthers. **LEAF** Leaf has 5–8 leaflets radiating from long leaf stalk (digitate). Leaflets taper to point at both ends, and are about the same length as long leaf stalk (petiole). **FRUIT** Cylindrical capsule reminiscent of a cucumber. Inside packed tightly with silky white floss that is commercially important. **HABITAT & DISTRIBUTION** Native to tropical America, Malaya, the Andamans and western India. Introduced to Sri Lanka. **ETYMOLOGY** *Ceiba* is a local name in Central America. *Pentandra* is a reference to five (*penta*) stamens.

Jam Tree ■ *Muntingia calabura*
(S: Jam Gaha)

STRUCTURE Small tree with canopy as wide as it is tall. Pale bark intricately patterned, on close inspection. Slightly drooping branches. **FLOWER** Each flower borne singly on stalk arising from leaf axil. More than one stalk can arise at a leaf axil. Five white sepals. Stigma has five lobes. Many stamens in centre of flower. **LEAF** Finely toothed leaf not symmetrical. Pointed at tip. Underside silvery and hairy. Upperside green and smooth. Leaves arranged in one plane on short stalks on opposite sides of a branch. **FRUIT** Tiny round berries with many seeds inside. Green fruits turn greenish-yellow to red when ripe. They are popular with birds, including the tiny Pale-billed Flowerpecker. **HABITAT & DISTRIBUTION** Native to tropical America, but widely planted in Asian countries. In Sri Lanka a common tree in home gardens and on roadsides. Fruit dispersed by birds. Has become established as a self-seeding tree. **ETYMOLOGY** Genus name honours A. Munting, a German physician and professor of botany. *Calabura* is a West Indian name.

Welang ▪ *Pterospermum suberifolium*
(S: Welang; T: Vilanku)

STRUCTURE Medium-sized tree. Sheds leaves seasonally (deciduous). Bark smooth and greyish. **FLOWER** Scented flowers yellowish-white on prominent stalks (peduncles) arising from leaf axils. Each flower stalk bears 1–3 flowers. Five narrow sepals. Calyx lobes pointed (acute). Five broad, whitish petals, shorter than sepals. Twenty stamens extrude prominently. **LEAF** Small leaves variable. They are sometimes long with roughly parallel sides (oblong), pinching abruptly at end to form point, and can also be slightly lobed in terminal half. Three to five prominent veins. Densely hairy on undersides. Distinct leaf stalks (petioles). Leaves arranged in a plane in two rows opposite each other (distichous). **FRUIT** Woody fruit narrow, pointed and covered in velvety white down. Within are winged seeds with wings longer than seeds. **HABITAT & DISTRIBUTION** Native to south-west India and Sri Lanka. **ETYMOLOGY** *Pterospermum* means winged seed, from the Greek *pteron* (wing) and *spermum* (seed). *Suberifolium* in Latin refers to having leaves like the Cork Tree.

Indian Tulip Tree ■ *Thespesia populnea*
(S: Suriya; T: Pu varathu)

STRUCTURE Small tree with many branches. Rough bark brown and knobbly. **FLOWER** Funnel-shaped flowers have red centres. Many stamens united into a tube in middle of flower. One or two large yellow flowers in leaf axils. Stalk (peduncle) shorter than leaf stalk (petiole). Sepals form bell shape. Petals yellow, turning mauve. Pollination is mainly by birds. **LEAF** Heart-shaped (chordate) leaves with pointed tips. Dark green on uppersides. Leaf stalk (petiole) up to 10cm long. Triangular, pointed leaf reminiscent of the Bo tree (see p. 95). Leaf has minute scales, especially underneath. **FRUIT** Smooth, rounded fruit, depressed at top centre. Five compartments contain silky seeds. **HABITAT & DISTRIBUTION** Common tree in the dry lowlands, especially on roadsides, where it is planted for its flowers and shade. In its natural state, favours areas near coasts. **ETYMOLOGY** *Populnea* means 'poplar-like'.

MORINGACEAE (HORSERADISH TREES)
This family of mainly trees and shrubs with a few herb species has a distribution from Africa, across the Arabian Peninsula, to southern India. It contains just one genus and 13 species. The tree trunks are swollen in a few species, and the plants often smell and taste of mustard. Their flowers are laterally symmetrical (zygomorphic), although in some species they tend towards being radially symmetrical (actinomorphic). They are often in axillary panicles, and are bisexual, with a tubular or cup-shaped receptacle and a nectary at the base. The five sepals are free and look like petals, and the five petals are usually white, yellow or red. They can be unequal is size and look like the flowers of a legume. The five stamens alternate with 3–5 sterile stamens (staminodes). The ovary is superior and on the end of a stalk (gynophore). The fruits are slender, beaked capsules that open with three valves.

Horseradish Tree or Drumstick ■ *Moringa oleifera*
(S: Murunga; T: Marunkai)

STRUCTURE Small tree with thin, pale trunk. Bark relatively smooth but can also be corky. Thrice-pinnate leaves give fern-like or feathery look to foliage. **FLOWER** Showy white flowers in axillary clusters. **LEAF** Lateral leaflets arranged on midrib in opposite pairs (pinnate), with rounded pinnae arranged in opposite pairs on these (bipinnate). Slender rachis thickened at base. Terminal pinna present. **FRUIT** Thin, long, three-angled green pods that turn brown on maturing. **HABITAT & DISTRIBUTION** Introduced from India. Widespread in gardens in the lowlands. Fast-growing tree that is very popular as its pods and leaves are consumed; also used as a live fence. Easily propagated by breaking off a branch and sticking it in the ground. Flowers popular with insects. **ETYMOLOGY** *Moringa* refers to a local name from Malabar, India. *Oleifera* is Latin for oil-bearing.

Caricaceae (Papayas)

The papayas comprise scrambling vines, shrubs and trees with a stem that is swollen and sometimes spiny. There are about 35 species in six genera. This is predominantly a family of the neotropics, at its most diverse in Mexico and the Andes. Most species have very prominent milky sap. Their leaves are alternate, palmately divided and spirally arranged at the ends of branches. In the Asian tropics, the family is familiar because of the widely planted introduced Papaya, which propagates naturally, dispersed by fruit-eating birds and mammals. Most species are unisexual, with male and female flowers on separate trees. The flowers are radially symmetric (actinomorphic), with five free sepals forming a cup. Male flowers have a nectary at the base with 10 stamens in two whorls. Female flowers have a superior ovary with five carpels.

Papaya ■ *Carica papaya*
(S: Papol, Gas Labu)

STRUCTURE Fast-growing small tree. Characteristic profile with crown of leaves on top of straight trunk heavily marked by leaf scars. Reminiscent of palms and cycads. Trunk soft and succulent, with milky sap, and typically unbranched. **FLOWER** Complex flower arrangement. Separate male and female flowers (unisexual) as well as bisexual flowers may be found on a tree. Sometimes male and female flowers are on different trees (dioecious), or a tree may have female and bisexual flowers. Male flowers have 10 stamens in two whorls. Stamens attached to petals. Female flowers usually have five rudimentary stamens, and bisexual flowers have five fertile stamens. Female and bisexual flowers borne in short-stalked clusters, and have ovaries surmounted by much-divided stigma. Male flowers borne in long, drooping panicles. Their five petals join together to form a tube with five lobes at the mouth; in female flowers, the five petals are hardly joined. **LEAF** Leaves form cluster at trunk top. Each leaf deeply divided into lobes, with up to seven lobes on a leaf. Lobes may in turn be subdivided. Leaf stalk long and hollow. **FRUIT** One of the best-known fruits in the world. Botanically a multi-seeded berry with seeds lining the walls of a single cavity. Fruit cylindrical, up to 35cm long, and green, turning yellowish or orange when ripe, and appearing close to tree's main stem. Black seeds are on fleshy pulp that is consumed by animals. Generally only trees with female or bisexual flowers bear fruit; however – rather strangely – male Papaya trees can at times bear fruit. **HABITAT & DISTRIBUTION** Native to tropical and subtropical America. **ETYMOLOGY** *Carica* is an old Latin name for the fruit of a fig tree (Papaya leaves resemble those of figs). *Papaya* is derived from a local name in the Caribbean.

SALVADORACEAE (TOOTHBRUSH-TREES)
This family of small trees and shrubs includes some species that are scramblers or climbers (scandent). They occur from tropical Africa, across Arabia, to Southeast Asia, near coasts where saline soils are found. Their simple leaves are thick and leathery, and arranged oppositely. The flowers are in axillary or terminal racemes of panicles (compound racemes). They can also form clusters at leaf axils (fasciculate). There are typically 2–5 sepals, fused to form a cup-shaped tube. The four petals (occasionally five) are free or fused at the base, which may have teeth lobes with glands on the inside. The ovary is superior, with two carpels. The short style is sometimes lobed into two.

Mustard Tree ■ *Salvadora persica*
(S: Maliththan; T: Uvay)

STRUCTURE Small tree with rounded appearance and drooping branches. Green foliage provides stark contrast to arid environments tree is found in. Pale bark. Old trees gnarled and full of character. **FLOWER** Greenish-white flowers in spreading racemes. Flowers arranged opposite each other on racemes. **LEAF** Oval leaves simple with smooth edges. Leaves and seeds have a mustard taste. **FRUIT** Red or dark purple on ripening, with single seed. Circular in shape but flattened somewhat (so not spherical). **HABITAT & DISTRIBUTION** Found from northern Africa, across Iran, to India. In Sri Lanka a common tree in the dry lowlands near the coast. Tolerant of saline soils and the dominant tree in coastal plains where the sea may make occasional incursions. Yala National Park is a good place to see many fine examples of the tree. **ETYMOLOGY** Genus name honours Jaume Salvador y Pedrol (1649–1740), a Spanish pharmacist and plant collector from Catalonia. *Persica* refers to its occurrence in Iran.

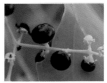

CAPPARACEAE (CAPERS)
Also referred to as the Capparidaceae, this family comprises about 650 species in around 17 genera of herbs, shrubs, lianas and trees from warm or arid areas. Mustard oils are produced from some species. Small trees in this family are among the most common trees in popular national parks such as Yala and Wilpattu. Their leaves are alternate, and one genus, *Apophyllum*, has no leaves and has photosynthetic stems. The leaves are typically simple. The inflorescences are variable, including racemes, panicles, corymbs and single flowers. The genus *Capparis* found in Sri Lanka contains species that are monoecious and dioecious.

Silver Caper ■ *Crateva adansonii*
(S: Lunuwarana; T: Navala)

STRUCTURE Elegant, medium-sized tree with distinctive pale or silvery, smooth trunk. Elephants seem to like using the tree as a rubbing post. **FLOWER** Bare of leaves when in flower. White-petalled flowers have many stamens and ovary borne on long, slender stalk (gynophore) extruding well out of corolla, making it a distinctive feature. Flowers have four sepals and four petals. **LEAF** Three pointed leaflets joined at leaf stalk make up leaf. Leaves have slight droop. **FRUIT** Rounded fruit fleshy with many seeds set within pulp. **HABITAT & DISTRIBUTION** Found throughout the dry lowlands. Absent in the arid zone. **ETYMOLOGY** Genus name honours Cratevas (1st century BC), a medicinal plant writer.

LECYTHIDACEAE (CANNONBALL TREES)
This family comprises shrubs and trees, with about 300 species in around 25 genera. They are mainly found in the neotropics, with some genera also occurring in Africa, and Asia to Australia. The family includes the well-known Brazil Nut. The plants' leaves are spirally arranged or are in two ranks. The simple leaves are large and usually without gland dots. They are clustered at the ends of bare branches. The inflorescences vary from racemes on the leaf axils, to panicles or long spikes. The flowering spike of the Cannonball Tree *Couroupita guianensis* is 1m long.

Fish Poison Tree ▪ *Barringtonia asiatica*
(S: Diya mudilla)

STRUCTURE Medium-sized tree; older trees attain heights of 30m. Bark greyish-brown, thinly fissured into rectangles and with tiny protuberances. **FLOWER** Numerous stamens joined together at base (connate) to form short, thick staminal tube. Petals pressed against (adnate) staminal tube. Anthers fixed at bases to filaments (basifixed). Long white stamens have pink tips. Flowering occurs at night and by morning the stamens and petals have fallen. **LEAF** Large, leathery leaves with distinct veins. Leaf scars clearly visible on twigs. **FRUIT** Distinct four-sided fruit separates this *Barringtonia* species from others. Fruit tapers to tip. Style and four sepals persist on leathery green fruit.

HABITAT & DISTRIBUTION Grows naturally in coastal areas from Madagascar to Australia (including Sri Lanka). However, many naturally growing trees in Sri Lanka may have been uprooted for coconut plantations. Planted inland as an ornamental tree. **ETYMOLOGY** Genus name honours Daines Barrington (1727–1800), an English naturalist. *Asiatica* refers to it being found in Asia.

SAPOTACEAE (SAPODILLAS)

This varied family comprising lianas, trees and shrubs, all containing typically white latex, is distributed across the tropics in humid forests, with a few species extending into arid regions and temperate latitudes. There are close to 1,300 species in around 58 genera. Some are spiny. Their leaves are simple and alternate, and the arrangement is usually spiral or in a plane, or sometimes the leaves can be in whorls. They have oppositely paired (pinnate) veins from the midrib. The flowers are axillary inflorescences, at times in panicle-like fascicles, which are on the bark (cauliflorous). The radially symmetric flowers (actinomorphic) are unisexual or bisexual. The sepal arrangement depends on the number of sepals: if there are 4–6, in a single whorl, 4–11 in a spiral, or 4–8 in two whorls. There are 4–18 petals, and 4–43 stamens. The stamens are fused at varying heights in the petal tube and are rarely free. The ovary is superior and the fruit is a berry or drupe, which may be fleshy, leathery or woody. The family is named after Sapodilla *Manilkara zapota*, originally cultivated in Mexico and now grown around the tropics for its edible fruit.

Mahua ▪ *Madhuca longifolia*
(S: Mi; T: Illupai)

STRUCTURE Tall tree with branches spreading out horizontally. Brownish or greyish bark deeply fissured and split into pieces. Milky sap. Deciduous, shedding leaves during dry season. **FLOWER** Flowers pale yellow and scented to attract bees (humans find flowers malodorous). They cluster on leafless shoots. There are about 16–24 anthers in whorls and the style projects out. Fleshy flowers rich in sugar, and can be cooked and eaten. **LEAF** Leaf long and tapers to blunt point. Young leaves rust coloured. Leaves clustered at ends of twigs. **FRUIT** Yellow fruit oval shaped and about 3cm long. **HABITAT & DISTRIBUTION** Occurs naturally in the dry lowlands, and possibly also in the wet lowlands, but as it is widely planted it is not clear if it occurs in the wet zone only as a result of introductions. **ETYMOLOGY** The Latin *longifolia* refers to the long leaf.

Ceylon Ironwood ■ *Manilkara hexandra*
(S: Palu)

STRUCTURE Prominent among trees in dry lowland national parks such as Yala and Wilpattu. Dominant member of climax vegetation in monsoon forests of the dry lowlands. Majestic tree, tall, with much-branched, spreading, lustrous green canopy. Shady canopy and big branches are a favourite of lounging young Leopards. When tree is in fruit, Sloth Bears gorge themselves on the fruit, then stagger about in an inebriated state. Bark dark, appearing black at times, and heavily corrugated into rectangular pieces. Tree's timber is valuable and much sought after. As a result, fine examples of the tree outside protected areas have been subject to illegal harvesting. **FLOWER** Small, pale yellow or white, scented flowers, occurring in ones and twos. Six or 12 stamens; sterile stamens (staminodes) may be zero, six or twelve alternating with stamens. Flower parts in multiples of six – possibly why the Greek word *hexandra* (for six) is part of the Latin name. **LEAF** Leaves dark green and stiff (coriaceous), and paler underneath. Leaf base tapers to acute angle. Tip rounded, sometimes indented. Leaf margins curl downwards. **FRUIT** Berries about the size of grapes, turning yellow (at times orange) as they ripen. Dark brown seed. Berry edible by humans and popular with a host of birds and mammals. **HABITAT & DISTRIBUTION** Native to Sri Lanka and India. In Sri Lanka a tree of the dry lowlands. Despite being so dominant, the only native tree in its genus, although a few other species in the same genus have been introduced (majority of species in genus are found in America). **ETYMOLOGY** *Manilkara* is derived from a local Indian name from Malabar.

EBENACEAE (EBONIES OR PERISIMMONS)

This family of deciduous and evergreen trees and shrubs is mainly found in the tropics, although a few species extend into temperate latitudes. The four genera comprise around 580 species, almost 90 per cent of them in the genus *Diospyros*. Their simple leaves are typically arranged alternately. The flowering arrangement is variable, and includes racemes and panicles. The radially symmetrical (actinomorphic) flowers are usually unisexual, but some species have bisexual flowers. The number of sepals is typically 3–5, but can vary up to eight. There are as many petals as sepals, which are fused into a tube. This is short in male flowers and shaped like an urn in female flowers. The number of stamens in a male can vary at 8–100, but more typically the number of stamens is a multiple of three or five times the number of petals. The fruits are usually berries, or sometimes capsules.

Ebony ▪ *Diospyros ebenum*
(S: Kaluwara; T: Karunkali)

STRUCTURE Medium-sized to large tree with very dark trunk – black in most trees, especially mature ones. This, together with the simple leaves, makes this a relatively easy tree to identify. Bark scaly. **FLOWER** Ebony trees can be monoecious (male and female unisexual flowers on same tree), dioecious (male flowers and female flowers on separate trees), or polygamous (unisexual and bisexual flowers on same tree). Male flowers in a cyme with 3–15 flowers. Female flowers solitary. **LEAF** Small leaves have nearly parallel sides before tapering to blunt tip. Midrib prominent. Leaves dark green on uppersides and paler below. Stiff leaf blade. **FRUIT** Fruit has 'collar' with four points. **HABITAT & DISTRIBUTION** Found in the wild in the dry lowlands. Some of the national parks (like Minneriya) have some fine old trees. Planted widely as an ornamental tree and for its highly prized wood. Native to Sri Lanka, India and Malaysia. **ETYMOLOGY** The Sinhala name Kaluwara means dark. The generic name *Diospyros* is derived from the Greek words *dios* (godly) and *puros* (grain of wheat). *Ebenum* is the Latin name for ebony.

PRIMULACEAE (PRIMROSES)
This family comprises about 1,000 species in 20 genera, familiar as cultivated plants for gardens. Most species are found in the northern hemisphere in temperate and mountainous regions. Only one occurs naturally south of the Equator, in South America. Some species are found in aquatic habitats.

Pink Ardisia ■ *Ardisia willisii*
(S: Balu Dan)

STRUCTURE Woody shrub that can form dense thickets near water. May grow to a height of 2m but most plants are much shorter. **FLOWER** Pink flowers borne in panicles. Flower stalk (pedicel) distinct, about 1cm long. Petals twisted and non-symmetrical. Stamens shorter than petals. **LEAF** Elliptic, simple leaves in clusters at ends of branches. Leaves smooth and shiny. Leaf tips blunt. **FRUIT** Fruit is a drupe with seeds having a fibrous coat. Fruits important for birds. **HABITAT & DISTRIBUTION** Endemic. Widely distributed in wet zone from the lowlands to the lower mid-hills. **ETYMOLOGY** *Ardisia* is derived from the Greek word *ardis*, referring to a point, in reference to a projecting style.

ERICACEAE (HEATHERS)
This widely distributed family of about 3,500 species in 125 genera, mainly growing as shrubs, but occasionally as trees, includes the small heathers of European heathland, as well as the shrubby evergreen rhododendrons of the Himalayas. The rhododendron found in Horton Plains grows as a shrub in exposed areas, and as a tree in sheltered parts near the cloud forest. South Africa is the centre of diversity for heathers, with rhododendrons having a centre of diversity in an area including western China, Tibet, Myanmar and Assam. These ericaceous plants generally prefer acidic soils and have a cosmopolitan distribution, but are absent in deserts.

Ceylon Rhododendron ▪ *Rhododendron arboreum zeylanicum*
(S: Ma Rathmal; T: Billi)

STRUCTURE Tree that can take the appearance of a small shrub, depending on where it is growing. In open grassland where the soil is poor and it is exposed to desiccating winds, grows as a small, stunted tree. Where conditions are more favourable in the company of other trees, grows as a medium-sized tree (it is unusual for a rhododendron to take the form of a tree, and other species grow as spreading bushes). Bark dark, cracked and spongy, demonstrating its fire-resistant qualities. **FLOWER** Clusters of red flowers densely packed together very striking, as in other rhododendron species. **LEAF** Leaf uppersides dark green and undersides silvery, with bristly hairs. **FRUIT** Fruit a hard, woody capsule. **HABITAT & DISTRIBUTION** Confined to the highlands at around 2,000m. Horton Plains National Park is the best location for seeing it. Unusual disjunct distribution, with other populations in the Nilgiris and Himalayas. Isolated populations on mountains in Sri Lanka and southern India believed to be relict populations after the retreat of the last ice age. Plant found in Sri Lanka considered a subspecies, *zeylanicum*, endemic to Sri Lanka. **ETYMOLOGY** From the Greek words *rhodo* (red) and *dendron* (tree), we have rhododendron, meaning red tree. *Arboreum* refers to being tree-like.

COFFEE

Rubiaceae (Coffee or Madders)

With more than 13,000 species in around 615 genera, this is the fourth largest family of flowering plants after the Orchidaceae, Asteraceae and Fabaceae. It is widespread and even found in the subpolar regions of the Arctic and Antarctic. The species diversity is greatest in the tropics and in humid tropical forests, and members of the family are often the predominant woody species in such forests. They grow as annual and perennial herbs, shrubs, lianas, epiphytes and trees. Some have chambers in the nodes for ants (like *Myrmecodia* species). Their simple, entire leaves are alternate or whorled. The inflorescences are variable, and the flowers are usually bisexual but also unisexual.

Citrus-leaved Morinda ▪ *Morinda citrifolia*
(S: Ahu; T: Nuna)

STRUCTURE Small tree or large, untidy shrub with pale bark. **FLOWER** Small white flowers clustered into spherical heads. Calyces of individual heads merged together to form fleshy mass. White petals of individual flowers separate, but joined basally to form tube within which are stamens. Calyx is green. In some calyces, a single leaf-like projection can grow that is distinctly longer than the petals. **LEAF** Shiny leaves in opposite pairs, with smooth margins. Elliptic in shape, tapering to point. **FRUIT** Green, fleshy mass comprises individual fruits, which are joined at bases. Each component fruit has a single seed. Composite fruit has short stalk arising from leaf axil. **HABITAT & DISTRIBUTION** Found in coastal areas in the lowlands. Planted in home gardens for its fruits and medicinal uses. **ETYMOLOGY** Generic name *Morinda* combines the Latin *morus* (mulberry) and *indus* (Indian), in reference to the fruit, which resembles a mulberry. *Citrifolia* refers to leaves resembling those of a citrus.

Oriental Bur-flower or Yellow Cheesewood
■ *Nauclea orientalis*
(S: Bakmi; T: Vammi)

STRUCTURE Tree generally medium sized, but can grow large in favourable conditions. Trunk straight and bark greyish-brown. In Colombo, a fine tree can be viewed from the far border of the Beddegana Wetland. **FLOWER** Scented orange or yellow flowers, growing in dense spherical heads. Flower heads have distinct stalk (peduncle) about 3cm long. Heads hang from stalk. **LEAF** Leaves can be 'fat oval' shape or heart shaped (cordate). They are in opposite pairs, and have hairy undersides. Distinct leaf stalk about 3cm long. **FRUIT**

Composite fruit, a fleshy mass containing fruits of individual flowers that were packed together in flower head. Contains many seeds. **HABITAT & DISTRIBUTION** Native to Malaya, Myanmar and Sri Lanka. Occasionally planted in gardens and as a street tree. **ETYMOLOGY** *Nauclea* in Greek means little ship. *Orientalis* in Latin means eastern.

LOGANIACEAE (LOGANIALES)
The loganiales are a diverse and widespread family with about 420 species in around 15 genera. They grow as annual and perennial herbs, woody climbers, lianas and large trees. Occupying a range of habitats, from arid areas to rainforests, they occur mainly in the tropics and subtropics, extending to warm temperate areas. In their opposite leaves, the petiole bases are joined by a raised line. The leaves are entire. The inflorescence is usually cymose and terminal, sometimes axillary. The flowers are radially symmetrical, and unisexual or bisexual. The corolla is generally fused to form a narrow tube with short lobes. In some genera the plants bear either female flowers or bisexual flowers (gynodioecious). The family is closely related to the gentians, and is divided into four tribes, one of which is the Strychnae, which contain poisonous alkaloids from which strychnine is made.

Strychnine Tree ■ *Strychnos nux-vomica*
(S: Goda Kaduru)

STRUCTURE Large to medium-sized tree with greyish bark and fairly thick trunk. **FLOWER** Five-petalled flowers with five stamens in tube formed by petals. Styles project beyond tube, which encloses stamens. **LEAF** Dark green leaves with three strongly marked nerves, arranged pinnately in opposite pairs with terminal leaflet. Leaf blunt tipped. **FRUIT** Large berry with hard rind. Enclosed within it and set within a soft pulp are several circular seeds that are flat and silvery. Seeds highly poisonous and used in the manufacture of strychnine. Poisonous to most animals, but some birds and langurs are able to break down the toxins and eat it. **HABITAT & DISTRIBUTION** Found in scrub and secondary forests in the dry lowlands. **ETYMOLOGY** In Latin, *nux-vomica* means ulcer-nut.

APOCYNACEAE (DOGBANES)
The dogbane family is predominant in the tropics and subtropics, although its range extends to temperate regions. In warm regions the family comprises mainly woody plants (trees, shrubs and lianas), but in temperate areas the plants are perennial herbs. Some of the shrubs are spiny. There are more than 2,000 species in 200 genera, including the familiar Oleander, which is extensively planted for its flowers. Many species are poisonous, being rich in alkaloids or glycosides.

Devil's Tree or Blackboard Alstonia ■ *Alstonia scholaris*
(S: Rukaththana; T: Elilaippalai)

STRUCTURE Tall tree that looks slim. Contains milky juice, as do other members of this family. Bark dark grey and rough. Acidic leaf litter results in bare understorey, and the seeming absence of other plants gave rise to the name Devil's Tree. **FLOWER** Small, greenish-white flowers borne on cymes with long stalks, massed at ends of twigs. At night flowers emit a sweet scent. **LEAF** Leaves arranged in whorls at ends of branches, with 3–7 leaves in each whorl. Leaves stiff and reflective on uppersides, with distinct two-toned appearance, being pale on undersides and dark green on uppersides. Tall, mature trees present a view of the undersides, giving the impression of pale foliage. Leaves long and blunt tipped. Short leaf stalks. **FRUIT** Long, slender green fruits like thin green pencils. They split to disperse several seeds that are flat and hairy. **HABITAT & DISTRIBUTION** Native tree with wide distribution, from Indian subcontinent, to Malay Peninsula and Australasia. **ETYMOLOGY** Genus name honours C. Alston (1685–1760), a Scottish botanist. *Scholaris* is a Latin reference to schools. Wood of tree was used to make school blackboards.

Wara ▪ *Calotropis gigantea*
(S: Wara; T: Manakkovi)

STRUCTURE Small shrub, typically about 1m in height, but can grow into form of a small tree. Bark very pale. Trunk and branches exude white sap that is harmful. Plant covered with soft down. **FLOWER** Flowers notable for their structure. Stamens form tube around style, and backs of anthers have large, fleshy appendages. Five petals are pointed. Closed petals on a bud do not overlap each other. Purple or white flowers are on cymes arising laterally. **LEAF** Fleshy, elliptic leaves positioned opposite each other. Leaf stalks almost absent. Leaf bases partially encircle stem. **FRUIT** Fruit comprises two fleshy green follicles. These split along the side to release seeds, which are flat with tuft of white hair at end. **HABITAT & DISTRIBUTION** Common plant in the dry lowlands, especially on

disturbed and open soils subjected to leaching of nutrients. **ETYMOLOGY** Genus name *Calotropis* is derived from the Greek *kalos* (beautiful) and *tropis* (ship's keel), a reference to the shape of the fruit. *Gigantea* is Latin for gigantic, or in this case being unusually high.

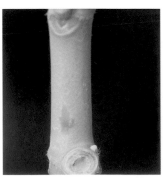

Poison Nut ■ *Cerbera odollam*
(S: Gon-kaduru; T: Nangi-ma)

STRUCTURE Small tree, neat and compact in appearance. When cut, exudes milky sap that is poisonous. **FLOWER** Prominent fragrant white flowers with five petals and five sepals. **LEAF** Light green leaves that are terminally clustered. Leaves long and narrow, typically ending in point. **FRUIT** Rounded, smooth green fruit that is fibrous and buoyant to aid dispersal by water. **HABITAT & DISTRIBUTION** Especially common in the lowlands adjoining water and in marshy areas. Used as a living fence in paddy fields because it thrives in wet soils. Wide distribution across Asia to Australia. **ETYMOLOGY** Genus name is derived from the Greek *Cerberus* (the hell-hound). *Odollam* is derived from a local Indian name.

Red Frangipani ■ *Plumeria rubra*
(S: Araliya)

STRUCTURE Grows as a small tree. Almost like a bush with low branches. Exudes milky white sap if bark is injured or one of the leaves is broken off. Bark greyish-white. Branches blunt ended, appearing to have been broken off and rounded. **FLOWER** Flowers in terminal cymes at ends of twigs. Small calyx has five segments. Five petals joined at bases to form very short tube, within which are stamens. Waxy petals then spread out. Flower stalk (petiole) about 5cm long. Flowers can be red or white, tinged with pink or purple, and have a lovely scent that is strongest at night. **LEAF** Leaves elliptic with small,

pointed tip. From central vein and almost at right angles to it are numerous parallel veins that run to leaf margin, which is smooth. Veins distinct. Leaves crowded at ends of branches. **FRUIT** Fruit consists of two long pods (follicles) joined at base. Each follicle splits along its length on one side along a suture. **HABITAT & DISTRIBUTION** Genus originates in tropical America. Tree introduced worldwide in the tropics as an ornamental for its attractive scented flowers. In Sri Lanka, a popular tree in temples, and the flowers are given as offerings. The residence of the Prime Minister is 'Temple Trees', a reference to the plumerias planted there. **ETYMOLOGY** Genus name honours Charles Plumier (1646–1706), a French botanist. *Rubra* is Latin for red.

White Frangipani ■ *Plumeria obtusa*

Almost all of what has been written about the Red
Frangipani (see opposite) also applies to this tree. Key
points for separating the species are flower colour and
leaf shape. In this tree, leaves are rounded, and some
may have the suggestion of a blunt tip. Flowers white
with yellow throat, and no hint of red or purple tinges.
ETYMOLOGY *Obtusa* refers to obtuse or blunt tips of
leaves.

Yellow Oleander ■ *Thevetia peruviana*
(S: Kaneru)

STRUCTURE Takes the form of a small tree or large bush. Branches close to the ground and foliage can be dense. Similar overall to the common garden plant, the Oleander. Milky white sap that is poisonous. Brown bark. **FLOWER** Bright yellow flowers in cymes. Flowers can also be white or pink. Calyx has five segments. Five petals joined to form funnel shape with five lobes at mouth. Five stamens are within petal tube (corolla). **LEAF** Long,

strap-like glossy leaves taper to points at both ends. Leaves densely arranged in spirals at ends of twigs. Leaf stalk very short, almost absent (subsessile). Smooth leaves shiny on uppersides. **FRUIT** Fruit somewhat triangular in shape, with elliptic cross-section. Green fruits turn black on ripening and typically contain four seeds, which are poisonous and can cause cardiac arrest. **HABITAT & DISTRIBUTION** Originated in the West Indies but now widely grown worldwide in the tropics. Easily grown from seeds and also planted as a hedge. **ETYMOLOGY** Genus name honours André Thevet (1502–1590), a French monk. *Peruviana* is a reference to originating in Peru.

Lamiaceae (Mints)
The Lamiaceae is a family of 239 genera with about 7,500 species worldwide, showing a huge variety in how they grow. It comprises five subfamilies, the Callicarpoideae, Prosthnatheroideae, Viticoideae, Tectonoideae and Lamioideae. Numerous familiar herbaceous plants, such as thyme, marjoram and lavender, are in this family, as is the large tropical hardwood tree, the Teak, cultivated extensively for its timber. Many of the herbaceous species are aromatic and have hairy leaves. The bisexual flowers are bilaterally symmetrical with five united sepals and five united petals. The petals are typically fused to form an upper lip and a lower lip, which gave rise to the old family name Labiatae (from the Latin word *labia* for lip). The flowers are usually in two clusters that superficially look like a whorl of flowers. The leaves are opposite in pairs at right angles to the previous pair.

Asiatic Beechberry ■ *Gmelina asiatica*
(S: Demata)

STRUCTURE Tall shrub that grows well above head height. Bark pale and fairly smooth. Branches have pairs of long thorns with small bases. Branchlets mat together to form dense canopy. One of the most common plants on roadsides in national parks of the dry lowlands. Similar G. *arborea* is not thorny and its leaves are about 7.5cm long (vs 4.5cm long in G. *asiatica*) with long tips. G. *asiatica* has rounded leaves. **FLOWER** Attractive single yellow flowers reminiscent of a trumpet. **LEAF** Leaves can be three lobed (most apparent in young leaves, see picture). **FRUIT** Small, rounded green fruit. **HABITAT & DISTRIBUTION** Common plant in the dry lowlands. On game drives in national parks such as Yala and Wilpattu, it can be seen beside safari roads. **ETYMOLOGY** Genus name honours J. G. Gmelin (1709–1755), a German botanist. *Asiatica* refers to it being found in Asia.

Teak ■ *Tectona grandis*
(S: Thekka; T: Tekku)

STRUCTURE Tall, large tree famed for its timber. Trunk straight with many branches spreading out laterally. Sheds leaves during dry season (deciduous). Smaller branches square in cross-section, and channelled. Enormous leaves elliptic, with prominent veins. Downy on underside. Bark light brown with thin strips flaking off. **FLOWER** White flowers borne on erect terminal panicles. Large inflorescence, up to 1m in height. Corolla five or six lobed. Petals fused into tube with five or six lobes. Style and five or six stamens

project beyond petals. **LEAF** Large elliptic leaves grow in opposite pairs. Short leaf stalks are four sided. **FRUIT** Fruit nearly spherical and more or less divided into four lobes. Typically contains one or two seeds. **HABITAT & DISTRIBUTION** Planted widely on roadsides and in public spaces and home gardens in the lowlands. Commercial plantations are in the dry lowlands. **ETYMOLOGY** *Tectona* is derived from the Greek *tekton* (carpenter). *Grandis* means large.

Horseshoe Vitex ■ *Vitex negundo*
(S: Nika; T: Vernochchi)

STRUCTURE Grows as a shrub or medium-sized tree. Branches relatively slender. Bark greyish-brown. Twigs and bases of flower stalks covered in white down. **FLOWER** Small violet flowers are on terminal panicles. Calyx has five teeth. Petals form tube at base but are separated beyond throat and end with five lobes. Lowest lobe is the largest. Four stamens projected (exserted) beyond petals. Throat of petal tube hairy. **LEAF** Each leaf divided into three or five leaflets radiating from common point on leaf stalk (digitate). Terminal leaflet is the largest. Lateral leaflets (when five leaflets are present) have very short, almost absent leaf stalks (subsessile). Leaves narrowly elliptic, taper to points at both ends and arranged in opposite pairs. Leaf margin can be smooth (entire) or slightly round toothed (crenate). Leaves generally smooth on uppersides with dense white down on undersides. **FRUIT** Egg-shaped, fleshy black fruit. Inside is a four-chambered stone with four seeds. **HABITAT & DISTRIBUTION** Grows naturally in the dry lowlands. Also planted in home gardens and on common land in villages. **ETYMOLOGY** Genus name is derived from the Latin name for a European shrub, *Vitex agnus-castus*. *Negundo* is derived from an old generic name for maples with divided leaves; a reference to the plant's divided leaves (a general trait of several species in this genus).

BIGNONIACEAE (BIGNONIAS)
The bignonias are found mainly in the tropics and subtropics, with species diversity being highest in the neotropics. There are about 850 species in around 110 genera, and the family includes climbers, lianas, shrubs and trees. They are adapted to a variety of habitats, from montane grassland to wet forests. Their leaves are compound and opposite with tendrils. The leaf arrangement is usually opposite, but can be alternate or in whorls. The flowers are solitary or in axillary or terminal racemes. The calyx has five fused petals, and the stamens are usually attached to the corolla tube.

Pink Tabebuia or Pink Tecoma ■ *Tabebuia rosea*

STRUCTURE Tall tree with greyish bark cleft into deep furrows in mature trees. Develops small buttress roots. In mature trees branching develops about 3–4m above the ground. Young trees have rough bark with numerous protuberances. **FLOWER** Showy, trumpet-shaped flowers with five-petalled lobes. Petals fall off easily, leaving behind long style. **LEAF** Leaves have 2–5 lance-shaped leaflets, each with a distinct stalk. In a leaf with, say, five lobes, two leaflets on sides will be stalkless and three central leaflets will have long stalks. Leaflets end with points. **FRUIT** Seed pods split (dehisce) to release seeds with thin, papery wings. **HABITAT & DISTRIBUTION** Native to central and northern South America. Popular decorative tree planted widely in the tropics. In Sri Lanka large trees can be seen in cities such as Colombo. **ETYMOLOGY** *Rosea* refers to the pink flowers.

Yellow Bells ■ *Tecoma stans*
(S: Kelani tissa)

STRUCTURE Small tree, at times like a bush. Many branched, with thick foliage. Bark light brown and corky. **FLOWER** Bright yellow flowers in terminal racemes. Calyx has five lobes. Petals form tube with five lobes. **LEAF** Leaflets lancelolate and with longish points and toothed margins. Leaflets in opposite pairs on midrib with terminal leaflet (imparipinnate). Terminal leaflet is the biggest. Leaflets stalkless (sessile). Flowers borne at ends of twigs. **FRUIT** Compressed, slender and long capsule, pointed at end, and about 10cm in length. Capsules green at first, turning brown when they ripen. **HABITAT & DISTRIBUTION** Native to South America, and planted widely in home gardens and on roadsides for its evergreen foliage and colourful flowers. **ETYMOLOGY** *Tecoma* is derived from a Mexican name. *Stans* is Latin for upright or erect.

African Tulip Tree ■ *Spathodea campanulata*
(S: Kudella Gaha)

STRUCTURE Tall tree with pale, somewhat greyish bark. Branches short, giving tree a slender profile. **FLOWER** Flowers clustered on terminal panicles. Orange or red petals about 10cm long and very striking. Inside, they can have yellow streaking. Their free ends have blunt tips and a narrow yellow margin. Four stamens project over petals, with brown anthers at ends. Calyx encloses flower in a bud and splits to allow flower to bloom. At this stage calyx hangs down to one side with tip curving back up again (recurved). **LEAF** Leaf has a number of opposite leaflets (pinnate) that are pointed, including a terminal leaflet (imparipinnate). Leaves up to 45cm long. Leaf margins smooth (entire). Leaflet stalks very short. **FRUIT** Thick, long, woody capsule about 35cm long. **HABITAT & DISTRIBUTION** Central African tree widely planted in tea estates in the hills, where it flourishes in the cool climate. It can be seen on roadsides and the grounds of estate bungalows. **ETYMOLOGY** *Spathodea* is derived from the Greek *spathe* (spoon or ladle), a reference to the shape of the calyx. *Campanula* is Latin for bell shaped, referring to tube formed by petals.

GOODENIACEAE (SEA-LETTUCES OR FANFLOWERS)
There are about 440 species in 12 genera in this family, with its centre being in Australia and New Guinea, the region that accounts for the significant bulk of genera and species. A few species, in the genus *Scaevola*, for example, have a worldwide distribution in the tropics. The plants' leaves are simple and spirally arranged. The flowers are in axillary clusters that are stalked. The flowers are bilaterally symmetric (zygomorphic). There are five sepals, and the five petals on the flower spread out palmately from a fat 'wrist'. Arising from the base ('the fat wrist') is a style that ends in a downturned, cup-shaped, densely hairy stigma. The flower looks as though it has had half of it lopped off. The pulpy fruit has a single seed, and the sepals are persistent at the terminal end of the fruit. The family name honours S. Goodenough (1743–1827), a botanist and bishop.

Sea-Lettuce Tree ■ *Scaevola taccada*
(S: Takkada)

STRUCTURE Common plant in seaside locations, taking the form of a shrub or small tree. General appearance is of dense, rounded foliage. **FLOWER** White flower appears to have half of it lopped off – unusual shape of flowers makes identification easy. Flowers form in cymes emerging from axils. **LEAF** Oval-shaped leaves in clusters at ends of branches. Leaf axils have tuft of white hair. **FRUIT** Oval fruit is a drupe. White or purple soft flesh surrounded by hard stone. **HABITAT & DISTRIBUTION** Found on coastline in Sri Lanka, and possibly most common in dry zone (though this may be more of a reflection of a lack of human disturbance). Widespread in Indo-Pacific. **ETYMOLOGY** *Scaevola* is a reference to the heroic Gaius Mucius Scaevola (500 BC), a Roman youth whose story is believed to be more mythical than historical. *Scaevola* means left-handed, the third name awarded to him after he thrust his left hand into a fire following a failed mission to kill the Clusian king laying siege to Rome.

Sea-lettuce Tree continued

ORDERS & FAMILIES IN BOOK

Order Cycadales
Cycadaceae (Sago)

Order Magnoliales
Myristicaceae (Nutmegs)
Annonaceae (Sweetsops & Soursops)

Order Laurales
Lauraceae (Bay-laurels)
Pandanaceae (Screwpines)
Arecaceae (Palms)

Order Zingiberales
Musaceae (Bananas)

Order Dilleniales
Dilleniaceae (Dillenias)

Order Celastrales
Celastraceae (Spindle Trees)

Order Oxalidales
Oxalidaceae (Wood Sorrels)

Order Malpighiales
Rhizophoraceae (Mangroves)
Clusiaceae (Mangosteens)
Salicaceae (Willows)
Euphorbiaceae (Spurges)
Linaceae (Flaxes)

Order Fabales
Fabaceae (Legumes)

Order Rosales
Rhamnaceae (Buckthorns)
Ulmaceae (Elms)
Moraceae (Mulberries, Jackfruits & Figs)

Order Myrtales
Combretaceae (Combretales)
Lythraceae (Pomegranates & Loosestrifes)
Myrtaceae (Myrtles, Eucalyptus & Cloves)
Melastomataceae (Melastomes)

Order Sapindales
Anacardiaceae (Cashews)
Burseraceae (Frankincense & Myrrh)
Sapindaceae (Lychee, Maples & Horse
    Chestnuts)
Rutaceae (Citruses)
Meliaceae (Mahoganies)

Order Malvales
Bixaceae (Annatoo)
Dipterocarpaceae (Dipterocarps or Marantis)
Malvaceae (Mallows)

Order Brassicales
Moringaceae (Horseradish Trees)
Caricaceae (Papayas)
Salvadoraceae (Toothbrush-trees)
Capparaceae (Capers)

Order Ericales
Lecythidaceae (Cannonball Trees)
Sapotaceae (Sapodillas)
Ebenaceae (Ebonies or Perisimmons)
Primulaceae (Primroses)
Ericaceae (Heathers)

Order Gentianales
Rubiaceae (Coffee or Madders)
Loganiaceae (Loganiales)
Apocynaceae (Dogbanes)

Order Lamiales
Lamiaceae (Mints)
Bignoniaceae (Bignonias)

Order Asterales
Goodeniaceae (Sea-Lettuces or Fanflowers)

Bibliography

Ashton, M. S., Gunatilleke, S., de Zoysa, N., et al. 1997. *A Field Guide to the Common Trees and Shrubs of Sri Lanka*. WHT Publication (Pvt) Ltd, Colombo.

Beentje, H. 2016. *The Kew Plant Glossary: An Illustrated Dictionary of Plant Terms*. Kew Publishing, Royal Botanic Gardens, Kew.

Benthall, A. P. 1946. *The Trees of Calcutta and its Neighbourhood*. Thacker Spink & Co.: Calcutta.

Blatter, E., Millard, W. S. & Stearn, W.T. 1954. *Some Beautiful Indian Trees* (2nd edn). Bombay Natural History Society: Bombay.

Christenhusz, M. J. M., Fay, M. F. & Chase, M. W. 2017. *Plant Families of the World*. Kew Publishing.

Corner, E. J. H. 1952. *Wayside Trees of Malaya*. Vol. 1. Authority, Singapore. Government Printing Office.

Corner, E. J. H. 1952, 1988. *Wayside Trees of Malaya*. Vol. 2 (3rd edn). Malaysian Nature Society: Kuala Lumpur.

Dassanayake, M. D. & Fosberg, F. R. (eds). *A Revised Handbook to the Flora of Ceylon*. Vols 1–7, published between 1980 and 1991.

Dassanayake, M. D., Fosberg, F. R. & Clayton, W. D. (eds). 1991–1996. *A Revised Handbook to the Flora of Ceylon*. Vols 8–10.

Dassanayake, M. D. & Clayton, W. D. (eds). 1996–2000. *A Revised Handbook to the Flora of Ceylon*. Vols 11–14.

Gardner, S. Sidisunthorn, P. & Anusarnsunthorn, V. 2007. *A Field Guide to Forest Trees of Northern Thailand*. Kobfai Publishing Project: Bangkok.

Heywood, V. H., Brummitt, R. K., Chulam, A. & Seberg, O. 2007. *Flowering Plant Families of the World*. Firefly Books.

Liyanage, S. 1997. *Sri Lanka's Mangroves*. Forest Department, Government Printing Press. Published in Sinhala.

Krishen, P. 2006. *Trees of Delhi: A Field Guide*. Dorling Kindersley India.

Mari Mut, J. A. 2017. Plant genera named after people (1753–1853). Published online by the author.

Miththapala, S. & Miththapala, P. A. 1998. *What Tree Is That? A Lay-person's Guide to Some Trees of Sri Lanka*. RukRakaganno: Colombo.

Pinto, L. 1986. *Mangroves of Sri Lanka*. Natural Resources, Energy and Science Authority of Sri Lanka.

Ratnayake, H. D. & Ekanayake, S. P. 1995. *Common Wayside Trees of Sri Lanka*. Royal Botanical Gardens, Peradeniya.

Stearn, W. T. 1983, reprinted 1991. *Botanical Latin: History, Grammar, Syntax, Terminology and Vocabulary* (3rd edn). David & Charles: Newton Abbott.

Tudge, C. 2005, reprinted 2006. *The Secret Life of Trees*. Penguin Books: UK.

Veevers-Carter, W. 1985. *Riches of the Rain Forest: Introduction to the Trees and Fruits of the Indonesian and Malaysian Rain Forests*. Oxford University Press: Southeast Asia.

Worthington, T. B. 1950. *Ceylon Trees*. The Colombo Apothecaries' Co. Ltd.

ORGANIZATIONS

The following list includes what may appear to be bird- or animal-centric organizations. However, their field meetings provide an excellent way to get out into good plant sites and meet people who know plants.

The Sri Lanka Natural History Society (SLNHS)
www.slnhs.lk, email: slnhs@lanka.ccom.lk
Founded in 1912, the SLNHS has remained an active, albeit small society with a core membership of enthusiasts and professionals in nature conservation. It organizes varied programmes of lectures for its members, and the subject matter embraces all fields of natural history, including marine life, birds, environmental issues and the recording thereof via photography and other means. It organizes regular field excursions, which include day trips as well as longer trips with one or more overnight stays.

Field Ornithology Group of Sri Lanka (FOGSL)
Department of Zoology, University of Colombo, Colombo 3. www.fogsl.net, email: fogsl@slt.lk
FOGSL is the Sri Lankan representative of BirdLife International, and is pursuing the goal of becoming a leading local organization for bird study, bird conservation and carrying the conservation message to the public. It has a programme of site visits and lectures throughout the year, and also produces the *Malkoha* newsletter and other occasional publications. Education is an important activity, and FOGSL uses school visits, exhibitions, workshops and conferences on bird study and conservation to promote its aims.

RukRakaganno, the Tree Society of Sri Lanka
http://rukrakaganno.wixsite.com/rukrakaganno, email: rukrakaganno09@gmail.com
RukRakaganno works with communities, particularly women, to care for and protect water resources and biodiversity. It is also concerned about trees in the urban environment, and currently manages the Popham-IFS Arboretum in Dambulla. Its publications include a newsletter for women in Sinhala, Tamil and English, and the booklet 'What Tree is That?' in the three languages. It also occasionally arranges tree walks.

Wildlife and Nature Protection Society (WNPS)
86 Rajamalwatta Road, Battaramulla. www.wnpssl.org. email: wnps@sltnet.lk
The WNPS publishes a biannual journal, *Loris* (in English) and *Warana* (in Sinhala). *Loris* carries a wide variety of articles, ranging from very casual, chatty pieces to poetry and technical articles. The society also has a reasonably stocked library on ecology and natural history. Various publications, including past copies of *Loris*, are on sale at its offices. The monthly lecture series is a sell-out success and if you need a seat you have to arrive early. It also organizes field trips.

The Young Zoologists' Association of Sri Lanka (YZA),
National Zoological Gardens, Dehiwala. www.yzasrilanka.lk, email: srilankayza@gmail.com

Facebook: YoungZoologistsAssociationofSriLanka. Twitter: @yzasrilankaYouTube: YZA Sri Lanka.
The YZA is a non-profit, non-governmental, voluntary youth organization based at the National Zoological Gardens, Dehiwala. The organization's focus is conservation through education. It meets at the zoo every Sunday at 2 p.m.

Forest Department
www.forestdept.gov.lk/index.php/en.
The Forest Department is the state institution tasked with the conservation and management of many of Sri Lanka's important forest reserves, including the lowland rainforests. It has a long track record of having excellent botanists on its staff and has issued a small number of publications. It also publishes a scientific journal, *The Forester*. These can be purchased from the head office in Colombo's suburbs.

Tour Operators

A strength of Sri Lanka lies in the presence of both general and specialist tour operators that can tailor a wildlife holiday. They include the following companies.

A. Baur & Co. (Travels), www.baurs.com
Adventure Birding, www.adventurebirding.lk
Birding Sri Lanka.com, www.birdingsrilanka.com
Birdwing Nature Holidays, www.birdwingnature.com
Bird and Wildlife Team, www.birdandwildlifeteam.com
Eco Team (Mahoora Tented Safaris) www.srilankaecotourism.com
Jetwing Eco Holidays, www.jetwingeco.com
Little Adventures, www.littleadventuressrilanka.com
Natural World Explorer, www.naturalworldexplorer.com
Nature Trails, www.naturetrails.lk
Walk with Jith, www.walkwithjith.com

Specific Acknowledgements

I had the advantage of growing up in Colombo in a house that was on half an acre, in which my mother had planted many native and well-established introduced plants. It was not until I was preparing this book that I realized how many plant species I had become acquainted with in my childhood thanks to my mother's gardening efforts. In the mid-2000s, Rohan Pethiyagoda made available to me a database of flowering plant species recorded in Sri Lanka. I have used this with modifications to help arrange species in the Angiosperm Phylogeny Group 4 (APG4) classification. The list has also been very useful in cataloguing and indexing my images. Rohan Pethiyagoda also founded the Wildlife Heritage Trust, which ushered a renaissance in biodiversity exploration and published a significant number of books. Among these is *A Field Guide to the Common Trees and Shrubs of Sri Lanka* by Mark Ashton et al. This was a useful reference when reviewing my images. I must confess, however, that when I lived in Sri Lanka, the absence of photographic illustrations made this a difficult book for someone like me who is not a botanist.

I have a large number of botanical books in my collection (in fact almost all the books listed in the Bibliography), which have, to one degree or another, influenced me. However, there a few publications that I need to specifically acknowledge. In writing the family accounts, I drew heavily on *Flowering Plant Families of the World*. Later on in the writing process, *Plant Families of the World*, which is an updated account using the results of the APG4, was also useful. The excellent *The Kew Plant Glossary* was a constant companion on my desk as I tried to decipher botanical English. For classification information, I used the online work of the APG4. A number of other books listed in the Bibliography were also consulted for the family accounts. Books I found especially useful include *The Trees of Calcutta and its Neighbourhood, Wayside Trees of Malaya, Ceylon Trees, A Field Guide to the Common Trees and Shrubs of Sri Lanka* and the multi-volume *A Revised Handbook to the Flora of Ceylon*. I am full of admiration for the work done by the authors of these more comprehensive books.

In the early 2000s, when I began to brand Sri Lanka as a destination for Leopard safaris and more generally for big game safaris, I was accompanied in the field by safari driver Amarasiri, who taught me the local names of species and helped me with the *Pictorial Guide to the Wildlife of the Dry Lowlands*, which was published in March 2006 by Jetwing Eco Holidays with funding from USAID. The tree photography for that book (available online on www.jetwingeco.com), laid the foundation for this book. Many of the Jetwing Eco Holidays guides have gone on to start their own companies or to become freelance guides. Some who accompanied me on field trips while preparing this book in the last few years (2016–2018) include Wicky Wickremesekera, Supurna and Suchithra Hettiarachchi (the famous Loku and Podi Hettis), Sam Caseer and Chandima Jayaweera.

On an ad hoc basis I consulted different people on the identification of specific images. I received help from Aruna Weerasooriya, Tharanga Wijayawickrama, Nadeera Weerasinghe, Darshani Singhalage and Nilantha Kodituwakku. Any mistakes in identification remain mine as no one was asked to review all the images I have used. My thanks also to the photographers who contributed images, which are individually acknowledged.

In the early 2000s, I developed a series of portable and affordable photographic guides to birds, butterflies and dragonflies with the Jetwing Eco Holidays team at a time when no one else was doing this. I believe these guides have been influential in encouraging the subsequent efforts by others in producing their own photographic guides, some of which are more comprehensive than the photographic field guides I produced to pave the way. I suspect the portable photographic guide approach brought to trees with this book will have the same outcome of encouraging others to attempt their own books in a similar format, and lead to a new generation of botanical explorers who will describe new species.

Between 2016 and 2018, the following people and companies hosted me to help me with photographing trees for this book: Chanaka and Ranitha Ellawala, Dallas Martenstyn at Dolphin Beach in Kalpitiya, Ajith Ratnayaka and his colleagues at Palmyrah House in Mannar and at the Backwaters Lodge in Eluvankulam, and Chitral Jayatilake and his team at Cinnamon Trails at Cinnamon Lodge. For a tree enthusiast, Cinnamon Lodge and Chaaya Village, which adjoin each other, are fabulous. Before this period, I have been in the field to photograph trees with Jetwing Eco Holidays, often on the grounds of Jetwing Hotels.

As already mentioned I find plants – and more specifically trees – a difficult subject. This book was photographed and written to help me much as it was done with others in mind. To help me get things right, the labelled trees at Beddegana Wetland Park and Diyasaru Park in Colombo, the Cinnamon Lodge in Habarana, the Popham Arboretum in Kandalama and the Henarathgoda Botanical Gardens were a useful resource. My thanks to all the people who, behind the scenes, were responsible for planting, labelling and nurturing these trees.

Tara Wikramanayake once again assisted with preliminary copy editing and posing questions. My thanks again to John Beaufoy and Rosemary Wilkinson for asking me to take on another book, and to Krystyna Mayer for her expert editing.

GENERAL ACKNOWLEDGEMENTS

Many people have over the years helped me in one way or another to become better acquainted with the natural history of Sri Lanka. My field work has also been supported by several tourism companies, as well as state agencies and their staff. To all of them, I am grateful. I must, however, make a special mention of the corporate and field staff of Jetwing Eco Holidays and its sister companies Jetwing Hotels and Jetwing Travels. During my 11 years of residence (as a working adult) in Sri Lanka, they hugely supported my efforts to draw attention to Sri Lanka as being super-rich in wildlife. In the corporate team, Hiran Cooray, Shiromal Cooray, Ruan Samarasinha, Raju Arasaratnam, Sanjiva Gautamadasa, Lalin de Mel and many others have supported my efforts. Past and present Jetwing Eco Holidays staff and Jetwing naturalists who have helped me include Chandrika Maelge, Amila Salgado, Ajanthan Shantiratnam, Paramie Perera, Nadeeshani Attanayake, Ganganath Weerasinghe, Riaz Cader, Ayanthi Samarajewa, Shehani Seneviratne, Aruni Hewage, Divya Martyn, L. S. de S Gunasekera, Chadraguptha Wickremesekera ('Wicky'), Supurna Hettiarachchi ('Loku Hetti'), Suchithra Hettiarachchi ('Podi Hetti'), Chaminda Jayaweera, Sam Caseer, Chandra Jayawardana, Nadeera Weerasinghe, Anoma Alagiyawadu, Hasantha Lokugamage ('Basha'), Wijaya Bandara, Suranga Wewegedara, Prashantha Paranagama, Nilantha Kodithuwakku, Dithya Angammana, Asitha Jayaratne, Lal de Silva, Chaminda Jayasekera and various interns. Chandrika Maelge, in addition to running the Jetwing Eco Holidays team, used her artistic and book-design skills to produce literature to brand Sri Lanka for wildlife, as well as to develop simple pictorial guides to build identification skills. Books such as this build on the work she did for the publications by Jetwing Eco Holidays, almost all of which are available online on its website. Many general managers and their teams at Jetwing Hotels supported my work.

My late Uncle Dodwell de Silva took me on Leopard safaris at the age of three and got me interested in birds. My late Aunt Vijitha de Silva and my sister Manouri got me my first cameras. My late parents Lakshmi and Dalton provided a lot of encouragement; perhaps they saw this as a good way of keeping me out of trouble! My sisters Indira, Manouri, Janani, Rukshan, Dileeni and Yasmin, and my brother Suraj, also encouraged my pursuit of natural history. In the UK, my sister Indira and her family always provided a home when I was bridging islands. Dushy and Marnie Ranetunge also helped me greatly on my return to the UK. My one-time neighbour Azly Nazeem, a group of then schoolboys including Jeevan William, Senaka Jayasuriya and Lester Perera, and my former scout master Mr Lokanathan, were a key influence in my school days.

My development as a writer is owed to many people – firstly, my mother Lakshmi, and more lately various editors and their colleagues, including the team at *Lanka Monthly Digest*, *Living*, the *Sunday Times* and *Hi Magazine*, who encouraged me to write. Various TV crews, especially the teams led by Asantha Sirimane at Vanguard and YATV, supported my efforts to popularize wildlife. My development as a naturalist has benefitted from the programme of events organized by the Wildlife and Nature Protection Society, Field Ornithology Group of Sri Lanka and Sri Lanka Natural History Society. In the UK, I have learnt a lot from the field meetings of the London Natural History Society and the London Wildlife Trust. I have also been fortunate to continue having good company in the field in Sri Lanka on my visits after I moved back to London. My friends and field companions include Ajith Ratnayaka, Nigel Forbes, Chitral Jayatilake and Ashan Seneviratne, who have arranged a number of field trips for me.

My wife Nirma and my two daughters Maya and Amali are part of the team. They put up with me not spending the time they deserve with them because I spend my private time working on the 'next book'. Nirma, at times with help from her parents Roland and Neela Silva, takes care of many things, allowing me more time to spend on taking natural history to a wider audience.

The list of people and organizations that have helped or influenced me is too long to mention individually, and the people mentioned here are only representative. My apologies to those whom I have not mentioned by name; your support did matter. Everything I know is what I have learnt from someone else.

REQUEST FOR IMAGES

If you would like to contribute images for a future edition of this book, please get in touch with me on gehan.desilva.w@gmail.com.

There are many species for which I need images of flowers or fruits, for example, to complete the coverage. I also generally request images for any of my books that can be improved with better images in future editions. The books I write and provide photographs for help to build capacity and improve the livelihoods of people on low incomes who live in areas adjoining parks and reserves. I am happy to use images from others who may wish to contribute to conservation. More details on contributing are on my blog on the following link (or else search for the blog wildlifewithgehan, then search for 'request for images'): http://wildlifewithgehan.blogspot.com/2018/05/request-for-images_13.